T207 Engineering: mechanics, materials, design

BLOCK 2

STATIC STRUCTURES: SHELTER AND PROTECTION

PARTS 3 AND 4

This publication forms part of an Open University course T207 *Engineering: mechanics, materials, design*. Details of this and other Open University courses can be obtained from the Student Registration and Enquiry Service, The Open University, PO Box 197, Milton Keynes MK7 6BJ, United Kingdom: tel. +44 (0)845 300 60 90, email general-enquiries@open.ac.uk

Alternatively, you may visit the Open University website at http://www.open.ac.uk where you can learn more about the wide range of courses and packs offered at all levels by The Open University.

To purchase a selection of Open University course materials visit http://www.ouw.co.uk, or contact Open University Worldwide, Walton Hall, Milton Keynes MK7 6AA, United Kingdom for a brochure: tel. +44 (0)1908 858793; fax +44 (0)1908 858787; email ouw-customer-services@open.ac.uk

The Open University
Walton Hall, Milton Keynes
MK7 6AA

First published 2003. Second edition 2006. Third edition 2009.

Copyright © 2003, 2006, 2009 The Open University.

All rights reserved. No part of this publication may be reproduced, stored in a retrieval system, transmitted or utilized in any form or by any means, electronic, mechanical, photocopying, recording or otherwise, without written permission from the publisher or a licence from the Copyright Licensing Agency Ltd. Details of such licences (for reprographic reproduction) may be obtained from the Copyright Licensing Agency Ltd, Saffron House, 6–10 Kirby Street, London EC1N 8TS; website http://www.cla.co.uk/

Open University course materials may also be made available in electronic formats for use by students of the University. All rights, including copyright and related rights and database rights, in electronic course materials and their contents are owned by or licensed to The Open University, or otherwise used by The Open University as permitted by applicable law.

In using electronic course materials and their contents you agree that your use will be solely for the purposes of following an Open University course of study or otherwise as licensed by The Open University or its assigns.

Except as permitted above you undertake not to copy, store in any medium (including electronic storage or use in a website), distribute, transmit or retransmit, broadcast, modify or show in public such electronic materials in whole or in part without the prior written consent of The Open University or in accordance with the Copyright, Designs and Patents Act 1988.

Edited and designed by The Open University.

Typeset in India by Alden Prepress Services, Chennai.

Printed and bound in the United Kingdom by Hobbs the Printers Limited, Brunel Road, Totton, Hampshire SO40 3WX.

ISBN 978 0 7492 2362 5

3.1

The paper used in this publication contains pulp sourced from forests independently certified to the Forest Stewardship Council® (FSC®) principles and criteria. Chain of custody certification allows the pulp from these forests to be tracked to the end use (see www.fsc-uk.org).

T207 presentation course team

Dr Jim Moffatt, course chair
Abi Lewis, course manager

T207 production course team

Professor Nicholas Braithwaite, course chair
Peta Jellis, course manager

Academic staff

Professor Adrian Demaid
Professor Chris Earl
Professor Lyndon Edwards
Mark Endean
Dr Michael Fitzpatrick
Dr Alec Goodyear
Dr Andrew Greasley
Dr Salih Gungor
Michael Hush
Jan Kowal
Dr Keith Martin
Dr Jim Moffatt
Dr Martin Rist
Dr Joe Rooney
Dr David Sharp
Graham Weaver
Dr George Weidmann

Support staff

Sylvan Bentley, picture researcher
Philippa Broadbent, materials procurement
Daphne Cross, materials procurement
Tony Duggan, learning projects manager
Vicky Eves, graphic artist
Garry Hammond, editor
Karen Lemmon, compositor
Vicki McCulloch, designer
Lara Mynors, media project manager
Lynn Short, software development

Part 3
Force and how to resist it

CONTENTS

1	Introduction	8
	1.1 Aims	8
	1.2 Resisting the force	8
2	Force and torque	10
	2.1 Combining forces	12
	2.2 Resolving forces into components	17
	2.3 The moment of a force	19
3	More about equilibrium	27
	3.1 Balanced forces	28
	3.2 Balanced torques	37
	3.3 Determinacy reprise	43
4	Summary	44
5	Learning outcomes	45
	Answers to exercises	46
	Answers to self-assessment questions	50
	Acknowledgements	51

1 INTRODUCTION

1.1 Aims

This part of Block 2 is about the forces that develop within a structure as a result of its self load and in response to applied loads. Its aims are:

- to examine how forces and torques can be separated into components and combined together;
- to revisit and extend the idea of equilibrium introduced earlier in the block;
- to use the concept of equilibrium in structural networks to establish the distribution of internal forces within simple static structures under given loading conditions.

1.2 Resisting the force

In Part 2 of this block I introduced you to the engineering problem of how to constrain motion in a network of rigid bodies interconnected at non-rigid joints, such as pin-joints and ball-joints. Knowing how to constrain motion enables engineers to design structural networks that can be used to solve the problem of supporting a roof or enclosing a space. In particular, such systems provide an efficient means of spanning large areas. Understanding the range of potential motions in a structure, and how to limit them, guides us to a successful configuration for the structure.

In order to specify the structure in terms of the dimensions of its members and the materials from which it should be made, engineers also require an understanding of the patterns and magnitudes of forces that can be generated within the structural network. With knowledge of these, engineers can design the structure to resist both self loads and externally applied loads. To do so they have to be confident that the internal forces, stresses and strains are neither too large in magnitude to cause failure (of the geometry or the materials) nor too small in magnitude to mean that the structure is inefficient (unnecessarily massive for the required loads, etc.). Such knowledge, coupled with a degree of imagination, leads to exciting solutions to the problem of providing shelter and protection (Figure 3.1).

You saw in Part 1 of this block that it is the loads on a structure or structural element that determine how it functions in practice, rather than the particular shapes of the components. Thus the primitive notions of pushing, pulling, shearing, bending and twisting are resisted by struts (in compression), ties (in tension), beams (in bending) and shafts (in torsion). In a real structural network all of these are present in general, but as a first approximation the design of such systems is often simplified by considering only tensile and compressive forces. The aim of this part of the block is therefore to investigate some of the patterns of tensile and compressive forces that can arise in structural networks. One of the key concepts that we shall use in our investigation is that of equilibrium – to function as a structural network, all the forces in a structure must be in equilibrium and Section 3 will focus on analysing structural networks through the concept of equilibrium. But first I am going to spend a little while introducing some of the ideas and mathematical techniques that you will need to use later.

PART 3 FORCE AND HOW TO RESIST IT

Figure 3.1 The Eurostar terminal at London's Waterloo Station, designed by Nicholas Grimshaw Partners and completed in 1993

2 FORCE AND TORQUE

Historically the concepts of 'force' and 'torque' arose in two different contexts. Firstly, the ancient Greeks investigated the subject we now refer to as 'statics'. They generally considered that objects in motion were in an 'erratic' or unnatural state, and that their proper preferred configuration was 'at rest' or static – that is, in a state of immobile equilibrium. It was Archimedes who studied equilibrium in considerable detail and one of the main subjects for which he is famous is the Law of the Lever – how it functions and how it works on the principle of balance. In modern terms this involves the equilibrium of forces and torques. Here I will mainly deal with the same sorts of static equilibrium situations that occupied much of Archimedes's time, but before I begin I will briefly mention the other context in which force and torque arise, namely 'dynamics'.

Almost 2000 years after the time of Archimedes, the first real attempt at studying objects in motion (rather than at rest) was begun by Galileo Galilei. Within a comparatively short time after this Newton had formulated his Laws of Motion, which essentially defined a force as the cause of a change in the motion of an object. You will study some of these dynamic systems later in the course, but here I want to draw your attention to the fact that Newton's approach led to a rethink of what is meant by 'force' in a system which is not changing its motion, such as the static systems studied by Archimedes. In brief, the problem is resolved by understanding that in a static system any forces that are present combine together in such a way that they balance each other – the effects of the forces cancel each other out. So it is as if there were no force acting on the system and it remains at rest. This is essentially what I mean by equilibrium – the forces balance each other. The forces are still present, but their *net* effect is zero.

But before I discuss (in Section 3) how to achieve equilibrium I shall explain some basic ideas about forces and their effects. I want you to develop an intuitive understanding of forces and torques and of how to combine them. I am going to start with a familiar everyday object – a shopping trolley – to which you can imagine applying forces. I choose a shopping trolley because you can easily see (or imagine) the effect a single force has on the trolley. It is then not too difficult to work out intuitively how we 'neutralize' the effect of one force by introducing a second (third, fourth, …) force to produce a combination of two (or more) forces on the trolley, which together bring it 'under control' – a state of equilibrium.

You can also model the behaviour of a shopping trolley by means of a book on a flat surface such as a desk or table. Applying forces to the short edge of the book roughly mimics the behaviour of pushing, pulling or twisting the bar at the back of a trolley.

Imagine you are in a supermarket with a shopping trolley. Most of us know instinctively how to move the trolley around from aisle to aisle and from display to display, and some of us do this apparently without thinking. But suppose you had to instruct a robot to move the trolley for you. What would the robot have to be able to do? Think about how you move the trolley. You push it, pull it, and turn it continually to keep it under control and avoid hitting the displays or other trolleys and people. The patterns of forces that are applied to the trolley by you (and the friction and reaction forces from the ground) are constantly changing. But although these patterns are complicated in detail there are really only a few different patterns that you need to consider in order to explain what is going on.

Typically you would guide the trolley with two hands placed on the bar, about shoulder width apart (Figure 3.2). This way of holding the trolley actually allows four main patterns of applying two forces to the trolley. Try the following exercise.

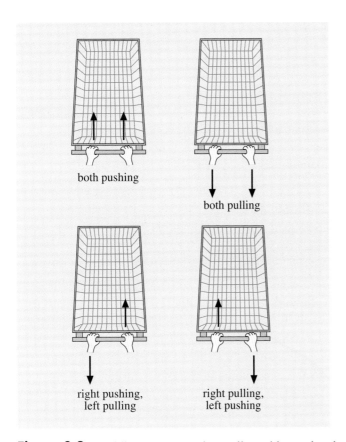

Figure 3.2 Guiding a supermarket trolley with two hands (viewed from above)

Exercise 3.1

Write down what happens when you do each of the following:

(a) push equally hard with both hands;

(b) pull equally hard with both hands;

(c) push with the right hand and pull equally hard with the left hand;

(d) push with the left hand and pull equally hard with the right hand.

So applying just two forces to the trolley can have four different effects, depending on the combination of pushing and pulling. You can see therefore that to understand how to achieve the different effects forces can have on a structure you need to know how forces behave in combination. And knowing how forces behave in combination is the key to understanding how equilibrium is achieved in a structural network.

Exercise 3.2

In no more than two minutes, think about and then write down what you would need to do to move the trolley forwards without turning it, by using just one hand.

You will recall from Part 1 of this block that you cannot adequately describe a force simply by saying 'I pushed with a force of 10 newtons'. A force has a *direction* associated with it – you push or pull towards or away from something. Moreover, a force is generally acting on an object at a point, such as the point at which your hand holds the trolley bar. So a force is described by three pieces of information:

- point of application
- direction
- magnitude.

One of the most important techniques used when working with forces is to combine these three aspects of a given force in a *graphical representation* of the system. Each force operating on the system is then represented by a straight line segment that:

- starts at the point of application;
- is aligned in the direction that the force is applied; and
- has a length that represents the magnitude of the force.

The line segment is also given an 'arrow-head' to indicate which way it is pointing. Such line segments therefore hold all the information necessary to characterize each force *vector*. In print, when I wish to indicate a vector I use bold-faced italic type. So, for instance two different forces might be labelled as $\boldsymbol{F}_1$ and $\boldsymbol{F}_2$. When I am concerned only with the *magnitude* of a vector, I use non-bold italic type, so the magnitudes of these two forces are F_1 and F_2. When writing vectors by hand, other conventions have to be used and you will find some guidance on this in the *Course Guide*.

Vectors are widely used to represent forces and many other directed quantities such as velocities and accelerations. But a word of caution is necessary since not all quantities with magnitude and direction can be represented by a vector. So, what else is needed besides a magnitude and a direction to be sure that I can represent something as a vector? Well, it turns out that the way quantities combine together is very important in establishing whether or not they are true vector quantities. So that's what we shall look at next.

> The term vector is also used in other non-mechanical contexts, such as computer graphics to describe shapes etc., and it is often used more loosely to describe a list of numbers or other entities in some computer applications. These contexts may have an association with some sort of notion of 'direction' but often they don't.

2.1 Combining forces

A force can be represented by a vector because two forces combine together in a certain way – they obey the so-called *parallelogram law*. To see how this works consider how you might use forces to lift an object and support it off the ground. You will need to pull vertically upwards with a force equal in magnitude to the weight of the object (actually with an initial force slightly greater than this in order to start the movement).

Now suppose the object is very heavy. You might need a helping hand to lift it. You could tie the middle of a rope to the object and with the help of another person you could lift the object by each of you pulling on one end of the rope. If you try this you will notice that neither you nor your helper needs to pull on the rope in a vertical direction (Figure 3.3).

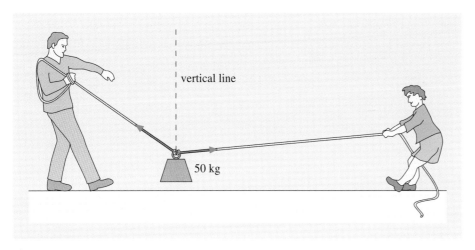

Figure 3.3 Two people supporting an object with a rope

The force you apply puts the rope in tension, and the vector representing this tensile force lies along the direction of the rope with length proportional to the magnitude of the force. The same is true of the force applied by your helper. The combined effect of the two tensile forces is a resultant force which counteracts the weight of the object so that it can be supported off the ground. You have achieved equilibrium by balancing three forces – the two tensions and the weight.

Note that the length of the rope is not very important here provided that it is long enough for you and your helper to hold. But the lengths of the vectors representing the forces depend on the strength of your pull and on the direction in which you pull. So, the lengths of the arrows in Figure 3.3 have nothing to do with the length of the rope – they represent the magnitudes of the forces. Hence the harder you pull on the rope the longer the vector representing the tensile force you are applying. The force vector might even be longer than the rope, as drawn to scale on a diagram. When analysing forces, I will draw vector arrows whose lengths represent the magnitudes of the forces, where they are known.

SAQ 3.1 (Outcome 3.1)

Thinking back to SAQ 1.10 in Part 1 of this block (and your own direct experience), if the ropes are both held at the same angle to the horizontal what can you say about the relationship between the angle and the amount of effort needed to support the weight?

Would you expect to be able to pull hard enough to make the rope horizontal as it supports the weight?

Imagine that you now perform an experiment to investigate the magnitudes and directions of the tensile forces required in the rope to support the weight. One way to do this is shown in Figure 3.4 (overleaf). Two separate 'ropes' are joined at a common point and the weight to be supported is hung from the join between the ropes. The two ropes are passed over pulleys and appropriate weights are attached to the ends so that the whole system is in equilibrium.

If I assume that there is no friction in the pulley system, each of the two extra weights provides a tensile force in its own rope equal to its own magnitude. This arrangement allows us to try different combinations of

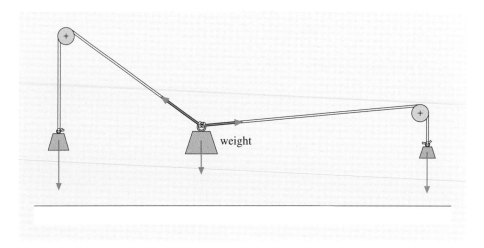

Figure 3.4 Two forces supporting a weight in a vertical plane

tensions and weights to be lifted, and thus see how the forces combine to produce equilibrium.

I can represent the two tensile forces by drawing two vectors pointing in the directions of the two forces, with lengths proportional to their respective magnitudes (Figure 3.5). The point of application of both forces is where the rope is connected to the weight, so each vector is directed away from this point.

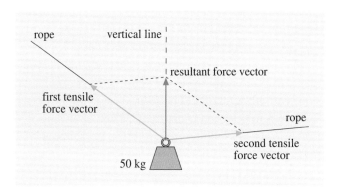

Figure 3.5 A parallelogram for adding two forces together

You can see now how the vectors can be thought of as forming the two adjacent sides of a parallelogram. What I find when I do this is that the diagonal of the parallelogram points vertically upwards when the weight is being held off the ground. What is more, its length is equal to the length of the vector representing the weight, whose vector naturally points vertically downwards.

When I try different combinations of weights I always find that the combinations that are in balance are the ones for which the diagonal of the parallelogram is vertical and equal in length to the magnitude of the weight supported. If the latter weight were to be supported by a single rope (rather than two ropes) then the tensile force in the single rope would have to be equal and opposite to the weight. So, when two ropes are used, the two tensile forces must combine to simulate a single rope and hence produce a vertical force equal and opposite to the weight. The diagonal of the parallelogram represents their combined effect, and this is referred to as their *resultant*.

SAQ 3.2 (Outcome 3.1)

Use the idea of a parallelogram of forces to explain why it is not possible to apply enough force to the two branches of the rope to support the weight with both ropes in a perfectly horizontal position.

So the *parallelogram law* allows us to combine two forces (vectors) by using them to form the sides of a parallelogram and then taking the diagonal as the resultant force. The force represented by the diagonal is considered to be the *sum* of the two individual forces represented by the sides. This can be justified by considering what the parallelogram would look like if the two forces retain their magnitudes but change their directions so that they point in the same direction. The parallelogram then collapses and its diagonal also lies in the same direction as the two forces. Its length equals the sum of the lengths of the sides. Hence when two forces are in the same direction their sum (resultant) has a magnitude equal to the sum of their magnitudes, and the vectors just add like numbers (Figure 3.6).

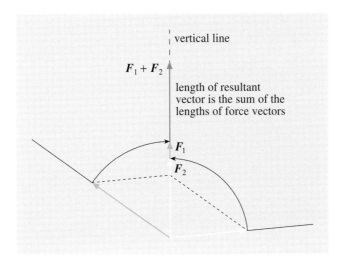

Figure 3.6 A collapsed parallelogram for adding two forces with the same direction

A diagram (such as Figure 3.5) showing the parallelogram law gives you a straightforward visual way to understand how two forces acting on an object add together. However, you must be careful to interpret any given parallelogram of forces properly. What the diagram is telling you is that any two given forces (the sides of the parallelogram) applied to an object can be *replaced* by a single force (their sum – the diagonal of the parallelogram). The important word here is 'replaced'. The diagonal of the parallelogram should be thought of as replacing the two sides. So, when you are considering the forces on an object, as soon as you add two of them using a parallelogram of forces, you should then forget the original two forces and only use the resultant diagonal force in their place. It's rather like adding numbers of things. For instance, you might have a bowl containing two apples and three oranges, so you would say that there are five pieces of fruit in total. But you would *not* say that there are two apples, three oranges *and* five pieces of fruit in the bowl. Don't make the mistake of thinking that the original two forces as well as the resultant diagonal force act on the object in combination.

Now that you have seen how to add two forces to get a resultant force you might be thinking that adding more than two forces becomes increasingly more complicated. However, the procedure is essentially the same and we don't need a new technique every time we add another force. Three or more forces are added by combining them two at a time. So if I have three forces *A*, *B* and *C*, I firstly combine *A* and *B* using the parallelogram law to get an intermediate resultant R_{AB}. I then combine R_{AB} with *C* to get a final resultant R_{ABC} of all three forces. This is shown in Figure 3.7.

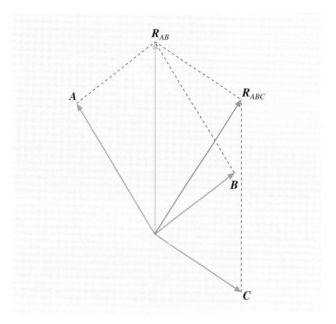

Figure 3.7 Combining three forces by repeated use of the parallelogram law

Exercise 3.3

Sketch the procedure for combining the three forces *A*, *B* and *C*, shown in Figure 3.7, to get a final resultant, by:

(a) firstly combining *B* with *C* to get an intermediate resultant R_{BC} and then combining R_{BC} with *A*;

(b) firstly combining *A* with *C* to get an intermediate resultant R_{AC} and then combining R_{AC} with *B*.

State how these two final resultants are related to each other and to the resultant shown in Figure 3.7.

I have introduced the parallelogram law and used it to show how to obtain the resultant (vector sum) of two or more forces. But it may also be used in the opposite fashion. By this I mean that instead of using it to combine (add) two forces together, it can also be used to split a given force into two other forces (Figure 3.8).

Hence a given force may be thought of as the diagonal of some parallelogram of forces and then be replaced by the two sides of the parallelogram. But, in this case, there are many possible parallelograms you could choose for a given force and all are equally valid. How do we go about deciding how to construct an appropriate parallelogram?

The answer is that you choose the most convenient one for the circumstances. It is a bit like taking a number, such as the population of the

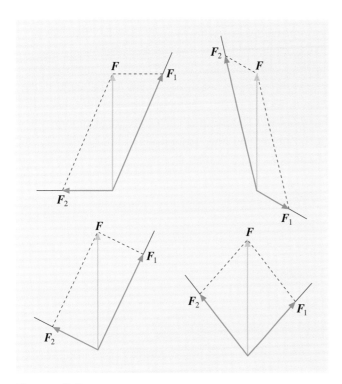

Figure 3.8 Four examples of using a parallelogram for splitting a force F into two component forces, F_1 and F_2

UK, and replacing it with two numbers, such as the number of males and the number of females. Alternatively, it could be replaced by the number of adults and the number of children, or the number of disabled people and the number of able-bodied people, and so on. In each case the total number of people (the population) is the same, but the way it is split into two numbers adding up to the total will be different.

The most common choice when separating a force into two components is to use a parallelogram that is a *rectangle*. This usually simplifies the situation considerably. Since the two component forces are at right angles to each other, you might recall from Part 1 of this block that they are *independent* of each other. Usually these may then be dealt with more easily than the original force (which they replace). This is essentially the process of resolving forces into components, and I will deal with this in more detail in Section 2.2.

You should note that the parallelogram law for combining forces is derived from experiment and not just from theory (remember Figure 3.4). It has been shown to be true in many different experimental situations. It provides a very useful graphical representation of the combined effect of two forces that are applied to the same object.

On the T207 CD-ROM you will find a useful software tool, 'Combining vectors', along with an introduction to its use.

2.2 Resolving forces into components

By now you might want to ask: 'Why would I split a given force into two components?'. The reason is that it is often difficult to do calculations using the original direction of the force, whereas it would be much easier to do calculations if the force were in some other direction such as the horizontal or vertical direction. By finding the components of the force, and working with these components instead of with the original force, the problem is usually simplified.

You have just seen how to replace a force with two other forces that have the equivalent effect by using a parallelogram. I suggested that the parallelogram of choice in most circumstances is a rectangle so that the force is replaced by two forces at right angles to each other and in the same plane as the original force, as shown in Figure 3.9. Components at right angles to each other are known as *orthogonal components* of the force. Although we usually choose our axes to be horizontal and vertical when analysing problems on or near the Earth's surface, we can choose our axes to be in other directions.

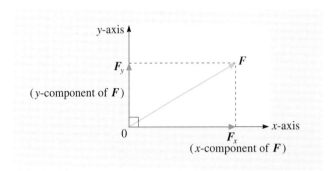

Figure 3.9 A force split into two components at right angles

Next we come to the question of establishing the magnitudes of the orthogonal force components. You should remember from Section 4.4 in Part 1 of this block that these have something to do with the magnitude of the original force and the angle between the force and one of the two components. I am now going to use the parallelogram law to demonstrate how this comes about.

In Figure 3.5 I showed you how to construct force components graphically. So, you could draw everything to scale and just measure the line segments representing their magnitudes. But in practice the accuracy of such measurements will depend on your drawing skills. A better method is to *calculate* the magnitudes of the horizontal and vertical components of the original force and this can be done using ▼The properties of right-angled triangles▲.

▼The properties of right-angled triangles▲

Take another look at the rectangle used in Figure 3.9 to obtain the horizontal and vertical components (say) of a force, F. The diagonal of the rectangle divides it into two right-angled triangles (Figure 3.10). Either of these can be used to calculate the magnitudes of the components of F. In each of the triangles (it usually doesn't matter which one you choose) the length of the hypotenuse represents F, the magnitude of F, and the lengths of the two sides represent the magnitudes F_x and F_y of the (horizontal and vertical) components. I know the length of the hypotenuse (it's just the magnitude of the original force), but I don't yet know the lengths of the sides. However, I do know something else, namely the angle θ between the direction of the original force and, say, the x-axis. I know this because that is how I specify the direction of F. This enables me to use trigonometry to find the lengths of the components of F.

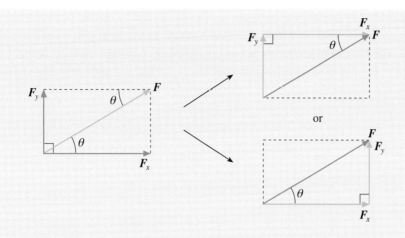

Figure 3.10 A force F and its two orthogonal components F_x and F_y form a right-angled triangle

Exercise 3.4

(a) Write out an expression for the magnitude F_x and an expression for the magnitude F_y in terms of F and the angle θ.

(b) If $F = 300$ kN and $\theta = 25°$, what are the magnitudes F_x and F_y?

So, I now know how to calculate the magnitudes of the horizontal and vertical components of a given force by using the cosine and sine of the angle between the direction of the given force and one of the axes. This technique is used widely in structural engineering.

2.3 The moment of a force

According to Newton, when I apply a force to a free particle (an object of negligible size) I cause the particle to accelerate in the direction of the force and the particle therefore changes its velocity in response, as you will see in Block 3. An extended free object (one whose size is not negligible) behaves in a similar way and its *centre of mass* accelerates in the direction of the force. But in this case there is usually another effect – you will know from your own real or imaginary experiments with shopping trolleys that applying a single force quite often causes an object to rotate about an axis.

The force will produce an angular acceleration, but I am not going to look at that aspect here. All I am interested in is the fact that a force produces a turning effect when acting on an extended object. This turning effect is referred to as a *torque*. It occurs because of the three characteristics of a force that I introduced earlier: point of application, direction and magnitude. Combining the direction with the point of application gives us a *line of action* of the force. The line of action is very important in determining the turning effect of a force because, unless the line of action passes exactly through the centre of mass of the object, there will be a turning effect.

Archimedes was very familiar with the idea of the turning effect of a force, arising from his study of the operation of levers. A *lever* is a device for changing the mechanical advantage, so that for example it enables a small force (the effort) to be magnified so that it can balance a large force (the load). It is essentially a beam pivoted at some point (the fulcrum) and it operates like a see-saw. We can classify levers as belonging to one of three

On the T207 CD-ROM you will find a useful software tool, 'Resolving vectors', along with an introduction to its use.

The centre of mass of an extended object may be defined to be that point through which any applied force does not produce a turning effect on the body.

different types, depending on whether or not the fulcrum lies between the effort and load, and on whether it is the effort or the load that is closer to the fulcrum (Figure 3.11). The crowbar would be an example of type (a) in Figure 3.11, the wheelbarrow exemplifies type (b), and tweezers exemplify type (c).

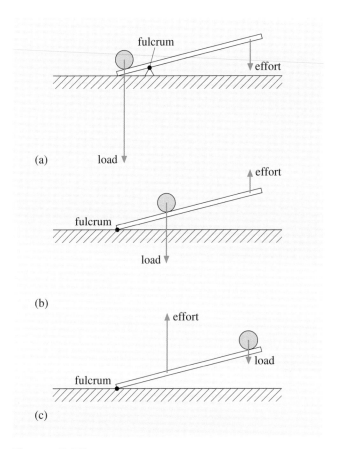

Figure 3.11 The three types of lever

For a given load, the further the effort is from the fulcrum the smaller it needs to be in order to balance the load. But strictly speaking it is the turning effects of the load and effort that are being balanced by the lever. Hence the lever balances torques, not forces. So how do we quantify a torque?

A force has a magnitude and a direction and it can be represented by a vector, as I have already explained. So what does a torque have? I will mainly look at two-dimensional (2D) situations here, because they are much easier to understand (and explain). The *magnitude*, M, of a torque in 2D has two parts. Firstly, the magnitude, F, of the force producing the torque contributes to the turning effect. But this cannot be the only factor because there is no turning effect if the line of action passes through the centre of mass of the object, as I have already suggested, regardless of the magnitude of the force.

So, the second factor affecting the magnitude of a torque is the position of the line of action of the force producing the torque. The further away the line of action is from the fulcrum or centre of the turning (the centre of mass in the case of a free object), the greater is the turning effect (hence burglars' interest in long crowbars). You can experience this effect for yourself if you try to push open a heavy door. Pushing near the outer edge (well away from the hinge axis) is much easier than pushing close to the hinge, because in the latter case the turning effect of the force you apply is quite small and you will struggle to rotate the door into an open position. The magnitude of a torque is therefore a combination of the *magnitude of the force* producing the torque and the *distance of the line of action* of the force from the fulcrum. But how do we combine these two factors to obtain the magnitude of the torque?

The magnitude M of a torque is defined as the magnitude, F, of the force multiplied by the shortest distance, d, from the centre of turning (the fulcrum) to the line of action of the force (Figure 3.12). This magnitude is known as the *moment* of the force about the point of rotation as was introduced in Part 1 of this block.

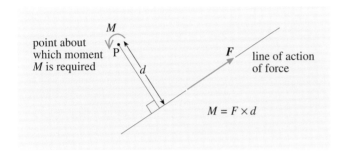

Figure 3.12 The moment M of a force F about the point P

Mathematically, the moment of the force in Figure 3.12 is:

$$M = F \times d$$

This tells me that I can produce a given moment by applying a large force along a line of action that is a small distance away from the fulcrum, or alternatively by applying a small force with a line of action that is a large distance away from the fulcrum. You can think of this in terms of the shopping trolley (Exercise 3.2). When using one hand the trolley turns more easily when your hand is near the outer end of the bar than when it is near the centre of the bar. This shows that you need to push harder when the distance of the line of action from the centre of mass is small, and conversely you need to push less hard when this distance is greater, in order to produce the same effect.

The trolley example is also instructive in another way. If you push with your left hand on the left side of the bar, you will turn the trolley in a clockwise sense as seen from above. Similarly, pushing with your right hand on the right side of the bar turns the trolley in an anticlockwise sense. If the force is the same and the distance is the same, in each case the magnitude of the torque will be the same. So, a torque with a given magnitude can produce two opposite effects – one a clockwise moment and the other an anticlockwise moment – depending on which side of the line of action the fulcrum is located. To specify a torque properly in two dimensions I must therefore give its magnitude and its sense (clockwise or anticlockwise). The situation is analogous to that of having a force with a given magnitude and line of action, but whose sense can be in either of the two directions along the line of action.

If I know the magnitude and direction of a force applied at some location in a structure, I know its line of action and hence I can calculate its moment about *any* point (and not necessarily just about some fixed point of rotation). This gives me a measure of how effective the force would be in producing a rotation about that point, even though in a structure an actual rotation is usually prevented by some constraint. As you saw earlier in the block, a knowledge of the effect an applied load will have on a structure in all respects is a vital aspect of designing the structure against failure.

Earlier, in Section 2.1, I showed you how to combine two forces together to get a new (resultant) force (and I also showed the opposite process, namely how to split a force into two components). There is a comparably simple procedure for ▼Combining two torques▲ to get a new torque.

▼Combining two torques▲

Suppose two forces F_1 and F_2, with different magnitudes and different lines of action, act on an object. Figure 3.13 shows this for two different cases.

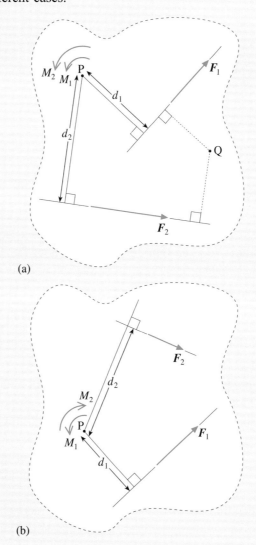

Figure 3.13 Combining two torques (two cases)

Choose any point P on the object and treat it as if it were a fulcrum about which the forces produce a turning effect. If F_1 and F_2 are the magnitudes of the forces, and d_1 and d_2 are the respective perpendicular distances from the fulcrum to their lines of action, then the magnitudes of the torques produced by the forces about P are respectively M_1 and M_2, where:

$$M_1 = F_1 \times d_1$$
$$M_2 = F_2 \times d_2$$

The combined turning effect of $\boldsymbol{F}_1$ and $\boldsymbol{F}_2$ about P is now obtained by adding the torques M_1 and M_2 to give a resultant torque M, where:

$$M = M_1 + M_2 = (F_1 \times d_1) + (F_2 \times d_2)$$

This is almost the whole story, except that the senses of M_1 and M_2 must be taken into account. I will adopt the convention that anticlockwise torques are treated as positive, and clockwise torques as negative. So in Figure 3.13(a) both M_1 and M_2 are positive, and the above resultant torque is correct. In Figure 3.13(b) however, M_1 should be treated as positive and M_2 as negative, so when they are added, the correct resultant torque is:

$$M = M_1 + M_2 = (F_1 \times d_1) - (F_2 \times d_2)$$

Note that if this calculation produces a positive value for M then the resultant torque acts in an anticlockwise sense about P. Conversely, if the value of M turns out to be negative then the resultant torque acts in a clockwise sense about P.

All of the torques that I have introduced so far are dependent on the point P that I chose as the fulcrum about which I took the moments of the forces. If I had chosen a different point, such as Q in Figure 3.13(a), I would have measured or calculated different values for d_1, d_2, M_1, M_2 and M. However, the magnitudes, F_1 and F_2, of the forces would have been the same as before. So, unlike the magnitudes of forces, the moments of torques depend on the point about which the moments are taken. This is where engineers can use their problem-solving skills to choose a convenient point about which to take moments. Often a point is chosen that lies on one or more lines of action of the forces. In this way the moments of these particular forces, about that chosen point, will be zero, and this simplifies the calculations.

A special case exists where the two forces are equal in magnitude and opposite in direction but have different lines of action, creating what is known as ▼A couple▲.

▼A couple▲

Suppose that two *equal and opposite* forces act on an object and I want to combine them to see what their resultant will be (Figure 3.14).

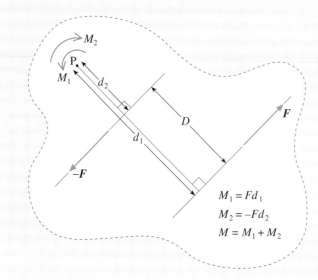

Figure 3.14 Combining two equal and opposite forces

If the forces act at the same point I can just form the parallelogram in the usual way and take the main diagonal as the resultant. But for two equal and opposite forces acting at a point the parallelogram collapses and the main diagonal has zero length and no defined direction. So it looks as if the two forces cancel each other, and since their lines of action are the same their moments about any point also cancel each other. Similarly, if two equal and opposite forces act at different points but their lines of action are the same, their moments about any point will also be equal and opposite. The overall effect in both cases is the same as if there are no forces at all on the object (remembering, of course, that the forces are still present).

However, if two equal and opposite forces are applied at different points and they have different lines of action, their magnitudes cancel, but their moments do not cancel. Instead they combine to give a non-zero torque.

Experimentally, it is always confirmed that a configuration of two equal and opposite forces with different lines of action have just a turning effect. This special configuration is known as a *couple*. You can try it for yourself with the shopping trolley. With your hands on the bar about shoulder width apart and placed symmetrically on either side of the centre-line, push forwards with your right hand and pull backwards with your left hand. If you push and pull with about the same magnitude you are applying a couple and the trolley should rotate anticlockwise without moving forwards or backwards.

Now let's see how to calculate the resultant torque produced by a couple – that is, two equal and opposite forces with different lines of action (Figure 3.14). The two forces each have magnitude F (remember they are equal and opposite, so their magnitudes must be equal) and I choose some point P about which I will take their moments (i.e. their turning effects). The moment of the first force, say, has magnitude $M_1 = F \times d_1$ tending to turn anticlockwise about the point P, where d_1 is the distance of P from the first line of action. The moment of the second force has magnitude $M_2 = F \times d_2$ tending to turn clockwise about the same point P, where d_2 is the distance of P from the second line of action (Figure 3.14). I can now combine these two moments algebraically by treating an anticlockwise moment as a positive magnitude and a clockwise moment as a negative magnitude. The resultant moment M of the couple is therefore given by:

$$M = M_1 + M_2 = (F \times d_1) - (F \times d_2)$$

This expression can be slightly simplified by noticing that the two terms on the right-hand side have the common factor F. The simpler expression is:

$$M = F(d_1 - d_2) = F \times D$$

where D is the perpendicular distance between the lines of action of the forces.

If I had chosen a different point, Q, about which to take the moments of the two forces, the distances d_1 and d_2 would be different numbers, but the last expression above shows that their difference would still be D (the distance between the lines of action). This shows that the magnitude of the resultant moment of a couple is always obtained by multiplying the magnitude of either of the forces by the distance between their lines of action. The point about which I take moments is irrelevant for a couple. In general the moment of a single force *does* depend on the point I choose for taking moments, but in the very special case of two equal and opposite forces with different lines of action (a couple) their combined moment is independent of the chosen point and only depends on how far apart are the lines of action.

If I want to apply a couple with a particular magnitude, M, to an object (so that I produce a specific turning effect with no linear movement), I can do this in many different but equivalent ways (Figure 3.15).

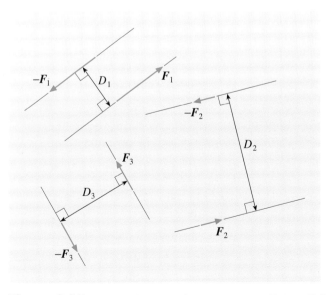

Figure 3.15 Equivalent couples: $M = F_1 D_1 = F_2 D_2 = F_3 D_3$

I can freely choose the magnitude, F, of the two equal and opposite forces (specifying how strongly I want to push and pull), and this determines the distance, D, between their lines of action (fixing how far apart I have to apply the two forces). Alternatively, I can freely choose the distance, D, and this fixes the magnitude of the forces, F, that I have to apply. For instance, two equal and opposite forces of 8 N magnitude, whose lines of action are separated by 3 m, produce the same couple as two equal and opposite forces of 6 N magnitude, whose lines of action are separated by 4 m. In the first case the magnitude of the couple is 8×3 N m, and in the second case it is 6×4 N m, both of which produce 24 N m of torque. Notice that I don't need to mention a direction for the two forces forming a couple – I only need to say that they have equal magnitude, *opposite* directions and have their lines of action separated by some fixed distance. The same couple is produced if the direction of the pair of forces is changed, for instance if they are rotated to any other orientation in their plane (think of a Catherine wheel) as shown in Figure 3.15.

Try applying a couple to the shopping trolley. Firstly place your hands close together near the centre of the bar, push forward with one hand and pull backward with the other hand with roughly equal strength. Secondly, place your hands on the extreme ends of the bar so that they are as far apart as possible. You should find that you have to push and pull quite strongly when your hands are close together on the bar to achieve the same turning effect as a much weaker push and pull when your hands are far apart on the bar. Also, try starting to push and pull the trolley from different starting orientations – you should find that the same turning effect is produced by the same strength of couple regardless of which way you and the trolley face at the start.

If you haven't already explored them, I suggest that you now try out the 'Combining vectors' and 'Resolving vectors' packages on the T207 CD-ROM.

3 MORE ABOUT EQUILIBRIUM

I used the idea of equilibrium in Part 1 of this block and again in the previous section of this part. I am going to explore it in yet more detail here, as it provides the basis for much of the analysis of static structures. For a system to be in (static) equilibrium the combination of all the forces must neither tend to move the system in a particular direction nor tend to rotate the system.

When a structural network functions correctly all parts of the system will be in equilibrium. This means that the forces will be balanced *at every point* in the network, and in particular at each joint. And it is not just the forces that balance. You have just seen how balanced forces can produce a couple, if applied at different points. So it is also the case that all the torques about any point on the network must also balance. This means that the network is able to resist the application of loads without changing shape or moving.

The following two conditions must therefore both be satisfied for a network to function as a structural network:

- **Equilibrium of forces**

    In a structural network the resultant of all of the forces in the network must be zero.

- **Equilibrium of torques**

    In a structural network the resultant of all of the torques about any point must be zero.

To achieve equilibrium in a network, any force acting on some part of the system must be balanced by at least one other force acting in the opposite direction on the same part of the system. Typically there will be more than two forces acting in several directions and mutually cancelling each other, but there must be *at least two* in order for that part of the system to be in equilibrium. Similarly, any tendency for a force to rotate a part of the system about some point must be balanced by at least one other force tending to rotate it in the opposite direction about the same point. In other words there must also be *at least two* torques cancelling each other.

These principles enable us to estimate the forces in each member of a typical load-bearing structure, such as that shown in Figure 3.16, of the sort that you will see in commercial buildings around the world. I shall come back to this example shortly, once I have developed a simple set of procedures out of the principles. To do so, I am going to use a more basic example.

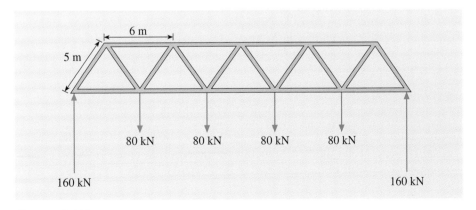

Figure 3.16 A model 19-member truss

3.1 Balanced forces

A familiar domestic situation in which forces must be balanced arises when I want to hang a picture on a wall. Suppose I use a cord attached symmetrically to two points on the top of the picture frame and I then place the midpoint of the cord over a hook fixed to the wall, as shown in Figure 3.17.

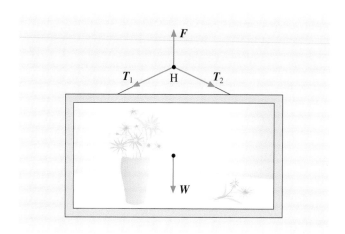

Figure 3.17 Hanging a picture frame from a hook, H

The forces on the hook at the point H are now in equilibrium (assuming the cord and the hook can support these forces) because the picture, cord and hook have effectively become a fixed structure in the plane of the wall. The *resultant* force on the hook must therefore be zero. But if I had to design the hook and the cord (or select suitable ones) I would need to know what the individual forces were in order to be sure the hook or cord wouldn't fail. What are these forces? The situation is similar to trying to find the forces on one of the joints in the structural network in Figure 3.16.

Remember that bold italic type indicates a vector, which carries information about both magnitude and direction.

Intuitively I know that the hook must support the weight W of the picture, and since the weight acts vertically downwards (directly towards the centre of the Earth), the hook must provide a reaction force F with the same magnitude as W but acting vertically upwards. In other words W and F are forces with equal magnitudes but opposite directions. Also intuitively, I know that the weight of the picture will pull the cord into tension and that, because of the symmetrical arrangement, the magnitudes of the two tensile forces T_1 and T_2 on either side of the hook will be equal. The three forces on the hook, F, T_1 and T_2, must have a resultant R which is zero because they are in equilibrium, so I can write:

$$F + T_1 + T_2 = R = 0$$

You can see how this works by using the parallelogram law twice to add F, T_1 and T_2 together as shown in Figure 3.18. Firstly add T_1 and T_2. Their resultant is a force equal to the weight W acting vertically downwards on the hook. When this (intermediate) resultant is added to the force F the final resultant is zero, since W and F have opposite directions and their magnitudes are equal, so the parallelogram collapses into a line.

The resultant of the forces is patently zero and I know their directions – vertical (F) and along the two branches of the cord (T_1 and T_2). Now I can develop ▼**General conditions for balanced forces acting at a**

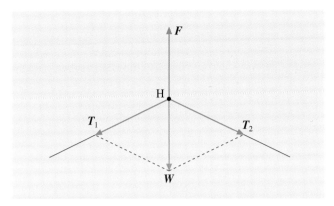

Figure 3.18 Combining the forces at the picture hook

point▲ and use these to calculate the forces acting on the picture hook. Notice that the outcome is a procedure that you have already used in Part 1 of this block. What I have done is to validate that procedure using the parallelogram law for combining vectors.

▼General conditions for balanced forces acting at a point▲

Remember from Section 2.2 that any force can be resolved into component forces in other directions and that in particular a given force can be resolved into two component forces in orthogonal directions, such as the horizontal and vertical directions (Figure 3.9). If I do this for each of the three forces, F, T_1 and T_2, acting at the picture hook, I will get:

$F = F_H + F_V$

$T_1 = T_{1H} + T_{1V}$

$T_2 = T_{2H} + T_{2V}$

This gives me three horizontal component forces, F_H, T_{1H} and T_{2H}, and three vertical component forces, F_V, T_{1V} and T_{2V}. I can now replace the sum of the original three forces with the sum of these six components, since they are equivalent to the original three forces, giving:

$F + T_1 + T_2 = (F_H + F_V) + (T_{1H} + T_{1V}) + (T_{2H} + T_{2V})$

The next step in the procedure is to rearrange this equation so that the three horizontal components are grouped together and the three vertical components are grouped together, giving:

$F + T_1 + T_2 = (F_H + T_{1H} + T_{2H}) + (F_V + T_{1V} + T_{2V})$

The horizontal and vertical forces combine to give resultant forces in each orthogonal direction, R_H and R_V respectively, such that:

$R_H = F_H + T_{1H} + T_{2H}$

$R_V = F_V + T_{1V} + T_{2V}$

The balance of forces on the hook is now:

$F + T_1 + T_2 = R_H + R_V$

The overall resultant force R is zero (because the system is in equilibrium), so:

$$R = F + T_1 + T_2 = R_H + R_V = 0$$

Since R_H and R_V are in orthogonal directions the only way they can add to zero is if each of them is zero. Hence for equilibrium:

$$R_H = F_H + T_{1H} + T_{2H} = 0$$

$$R_V = F_V + T_{1V} + T_{2V} = 0$$

These two conditions are vector equations, one relating the horizontal components and the other relating the vertical components. But all the horizontal components are in the same direction (that is horizontal), and the same is true of the vertical components – they are all in the vertical direction. Remember from Section 2.1 that adding forces along the same direction is just like adding numbers (scalars). The parallelogram for each addition of two horizontal components has collapsed into a line (see Figure 3.6), and similarly for the addition of two vertical components. So the above two vector equations for horizontal and vertical components can be rewritten as scalar (that is, non-vector) equations, but we need to be careful about how we combine the various terms. We adopt a convention that, in the scalar equations, horizontal components directed to the right are positive and those to the left are negative. Similarly, vertical components directed upwards are positive, whereas those directed downwards are negative.

- The net horizontal component is the sum of all the magnitudes of those components directed to the right, less the total of the magnitudes of all those directed to the left.

- The net vertical component is the sum of all the magnitudes of those components directed upwards, less the total of the magnitudes of all those directed downwards.

For the equilibrium of forces, the net horizontal component must be zero, and the net vertical component must also be zero. In other words, for the equilibrium of forces in two dimensions, two conditions (equations) are required to relate these forces together.

Now I am in a position to work out the magnitudes of the reaction force at the hook and the tensions in the cords. If you look back to Figure 3.18 you will now see that the two scalar conditions for equilibrium of forces in 2D become:

$$R_H = 0 - T_{1H} + T_{2H} = 0$$

$$R_V = F_V - T_{1V} - T_{2V} = 0$$

Note that F_H is zero since F has only a vertical component.

Example

A typical geometry for the hanging picture is shown in Figure 3.19.

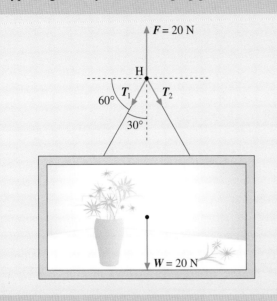

Figure 3.19 A picture frame with the cord forming an equilateral triangle

Take the weight of the picture and its frame to be 20 N and assume the cord is attached so that it forms an equilateral triangle. The angle that each branch of the cord makes with the horizontal direction is therefore 60° (30° to the vertical direction), and so the branches are arranged symmetrically on either side of the vertical. For this reason the magnitudes (T_1 and T_2) of the tensile forces in each branch should be equal, so let's say $T_1 = T_2 = T$. Resolving the forces in the vertical direction, I get the following equation for the vertical components:

$$F_V - T_{1V} - T_{2V} = 20 \text{ N} - T_1 \cos 30° - T_2 \cos 30°$$
$$= 20 \text{ N} - T \cos 30° - T \cos 30° = 0$$

Simplifying this I get:

$$20 \text{ N} - 2T \cos 30° = 0$$

and rearranging this I have the following equation for T:

$$T = \frac{20 \text{ N}}{2 \cos 30°} = 11.5 \text{ N}$$

This is the magnitude of the tensile force in each branch of the cord.

If I also resolve the forces in the horizontal direction, I get:

$$T_{1H} - T_{2H} = T_1 \cos 60° - T_2 \cos 60° = T \cos 60° - T \cos 60° = 0$$

This does not give any more information, other than to confirm that the horizontal components are equal and opposite, and hence cancel.

Now I want you to see what happens if the weight of the picture remains 20 N, but the cord is attached differently to the picture frame so that it forms an isosceles triangle, with each cord branch making an angle of 45° to the horizontal, as shown in Figure 3.20. The situation is still a symmetrical one, with the two tensile forces being equal to each other in magnitude. They are now obviously different in direction from what they were before, but they also must be different in magnitude.

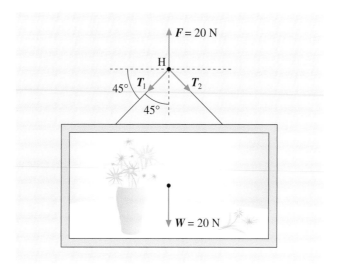

Figure 3.20 A picture frame with the cord forming an isosceles triangle

Exercise 3.5

Calculate the value of T for the arrangement shown in Figure 3.20.

Notice that the new force is bigger in magnitude than the one I obtained when the cord formed an equilateral triangle, so the cord has to support a greater tensile force when it is attached at a 'shallower' angle to the horizontal – the shallower the angle, the greater the magnitude of the force. In one extreme case the cord could be attached 'steeply', so that both branches are vertical (there would then be only one attachment point to the frame). For this case you should expect the tensile force to have its smallest value. Resolving vertically for this case, I get:

$$F_V - T_{1V} - T_{2V} = 20 \text{ N} - T_1 \cos 0° - T_2 \cos 0°$$
$$= 20 \text{ N} - T \cos 0° - T \cos 0° = 0$$

Simplifying this I get:

$$20 \text{ N} - 2T \cos 0° = 0$$

and rearranging this I have the following equation for T:

$$T = \frac{20 \text{ N}}{2 \cos 0°} = \frac{20 \text{ N}}{2} = 10 \text{ N}$$

You should not be surprised to see that the tensile force T in each of the two cord branches now has a magnitude of 10 N, equal to half the weight of the picture frame.

In Figure 3.21 I have summarized graphically how (for a given fixed weight of picture frame, hung symmetrically) the steepness of the cord affects the magnitude of the tensile force. In all cases the vertical component of the tensile force must be the same (equal to half the weight).

I am now in a position to return to the truss shown in Figure 3.16 and attempt to estimate the forces in each of its members. To do so involves making one crucial approximation: to model the truss as a network of rigid links connected by revolute joints – a 'pin-jointed frame'. ▼**Solving the truss**▲ to arrive at values for the force in each member of the truss can then be done by a series of very simple calculations based entirely on the conditions for vertical and horizontal equilibrium.

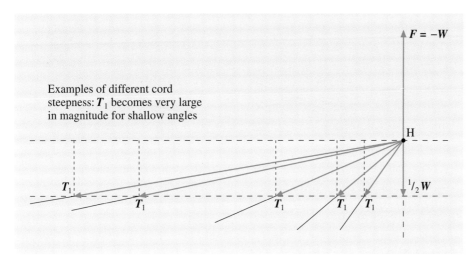

Examples of different cord steepness: T_1 becomes very large in magnitude for shallow angles

Figure 3.21 How the magnitude of the tensile force varies with cord steepness

▼Solving the truss▲

The model of the 19-member truss is as shown in Figure 3.22(a).

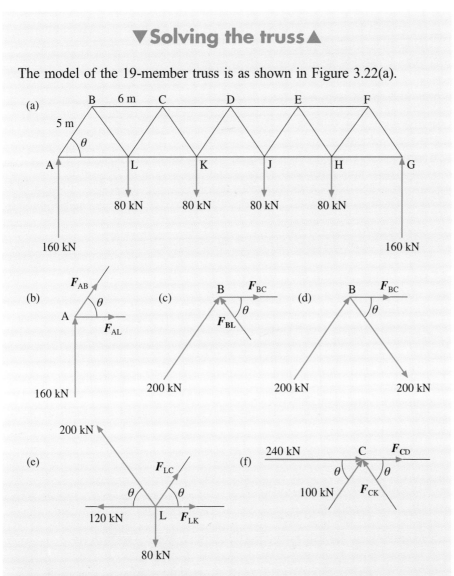

Figure 3.22 A 19-member truss

The applied loads (live and/or dead) have been simplified to a point load of 80 kN applied to each lower joint of the truss, not including the end joints on which the truss is supported. Truss and loads are symmetrical, so calculations for only half the truss need be carried out. The joints of the truss are labelled alphabetically as indicated and the force in each link is given a subscript indicating which two joints are connected by the link.

Note that the geometry of the truss is based on a series of triangles each having a base of 6 m and sides of 5 m. The vertical height of each triangle is then 4 m giving $\cos\theta = 3/5$ and $\sin\theta = 4/5$, which means that calculations can be done entirely without an electronic calculator.

As the following calculations progress, label each of the links in the truss with a 'T' if the force in the link emerges as tensile and 'C' if it is compressive.

Vertical equilibrium for the truss as a whole

Working with a symmetrically loaded truss allows me to establish the reaction forces at the supports using just vertical equilibrium. (You will see shortly how to deal with situations where this is not possible.)

The applied loads are balanced by equal reaction forces at each support. The applied loads sum to $4 \times 80 = 320$ kN, so each reaction force is 320/2 or 160 kN acting vertically upwards.

Equilibrium at joint A (Figure 3.22b)

A free-body diagram for joint A is shown in Figure 3.22(b). Remember, the sign convention dictates that the unknown forces are drawn as positive – upwards and to the right. F_{AB} is resolved into a vertical component, $F_{AB}\sin\theta$, and a horizontal component, $F_{AB}\cos\theta$.

For vertical equilibrium:

$$F_{AB}\sin\theta + 160 \text{ kN} = 0$$

or

$$F_{AB} = \frac{-160}{\sin\theta} \text{ kN}$$
$$= -160 \times \frac{5}{4} \text{ kN}$$
$$= -200 \text{ kN}$$

So the force in link AB acts downwards on A and is therefore a compressive force of 200 kN.

Horizontally:

$$F_{AL} + F_{AB}\cos\theta = 0$$

or

$$F_{AL} = -F_{AB}\cos\theta$$

Substituting in the value for F_{AB} that I have just calculated gives:

$$F_{AL} = 200 \times \frac{3}{5} \text{ kN}$$
$$= 120 \text{ kN}$$

The force in link AL is therefore tensile, acting to the right, away from A.

Equilibrium at joint B

Figure 3.22(c) is a free-body diagram of joint B. As I now have a value for the force in link AB, I have put this directly onto the diagram. This approach reduces the number of algebraic expressions in the subsequent equations. It is also important to realize that the compressive force in AB acts *upwards* on joint B and is therefore positive in the next calculation. This upward force of 200 kN is now resolved into horizontal and vertical components to find the forces in BC and BL.

Vertically:

$$F_{BL} \sin \theta + (200 \times \sin \theta) \text{ kN} = 0$$

or

$$F_{BL} \sin \theta = -(200 \times \sin \theta) \text{ kN}$$

so

$$F_{BL} = -200 \text{ kN}$$

The force in BL is acting downwards from B and is therefore tensile. I need to use the value of F_{BL} in the next calculation to find the force in BC. It's least confusing to draw yet another free-body diagram of joint B with all the known values for the forces acting on the joint shown in their correct directions. This is Figure 3.22(d).

Horizontally:

$$F_{BC} + (200 \times \cos \theta) \text{ kN} + (200 \times \cos \theta) \text{ kN} = 0$$

or

$$F_{BC} = -(400 \times \cos \theta) \text{ kN}$$

so

$$F_{BC} = -240 \text{ kN}$$

Equilibrium at joint L (Figure 3.22e)

Four links join at L and there is an applied load, but there is only one unknown force acting vertically, so again I can establish values for the unknown internal forces. I must remember, though, that the force in BL acts away from L and its vertical component is therefore positive in the calculation.

Vertically:

$$F_{LC} \sin \theta + (200 \times \sin \theta) \text{ kN} - 80 \text{ kN} = 0$$

or

$$F_{LC} \sin \theta + 160 \text{ kN} - 80 \text{ kN} = 0$$

Therefore,

$$F_{LC} \sin \theta = -80 \text{ kN}$$

so

$$F_{LC} = -100 \text{ kN}$$

The force in LC is acting downwards on L – a compressive force. I shall not draw yet another free-body diagram for joint L for the next calculation. I suggest you do this yourself, or add the value for F_{LC} on to Figure 3.22(e) acting in the correct direction. The following calculation involves three known forces, two of which need to be resolved horizontally.

Horizontally:

$$F_{LK} - (100 \times \cos \theta) \text{ kN} - (200 \times \cos \theta) \text{ kN} - 120 \text{ kN} = 0$$

or

$$F_{LK} - 60 \text{ kN} - 120 \text{ kN} - 120 \text{ kN} = 0$$

so

$$F_{LK} = 300 \text{ kN}$$

Exercise 3.6
By balancing the forces acting on joint C, calculate the values of the forces in CK and CD. Figure 3.22(f) is a free-body diagram for joint C with the currently known forces included.

Exercise 3.7
Draw a free-body diagram for joint K including all the known forces (don't forget the applied load).

Calculate the values for the forces in KD and KJ. Take care, as the results may surprise you.

Once you have determined the forces in each of the members of the truss, some interesting patterns result that tell us a lot about using structural frameworks as load-bearing parts of buildings. What I want you to do next is to draw together what we have just established about the truss and to compare it to analyses that we undertook in Part 1 of this block.

Exercise 3.8
Sketch the 19-member truss from Figure 3.22 with its applied loads and write next to each link the value of the internal force we have calculated for that link. Add a 'T' or a 'C' depending on whether it is in tension or compression, respectively.

The truss could be replaced by a beam carrying a uniformly distributed load (UDL). In line with a sketch of the truss, draw the outline of the shear force distribution and bending moment distribution in a beam carrying a UDL (refer back to Figure 1.27 in Part 1 if you don't recall what these are).

Clearly there is close correlation between the behaviour of a beam and the forces developed in a truss under similar loading conditions. I want you to think back to Section 4.1 in Part 1 of this block, and in particular to SAQ 1.6. There you were asked to think about the use of I-section beams to support roofs and floors in buildings. Can you see the similarity between the truss and an I-section beam?

I like to think of a truss as an I-section beam with surplus material removed from the web joining the upper and lower surfaces, leaving just a network of struts and ties that maintain the relative positions of the two surfaces. Of course, to go the last step and work out the optimum size and shape of each link in the network calls for information about the strength and stiffness of the materials to be used, and we shall have to leave that until later.

SAQ 3.3 (Outcomes 3.3 and 3.4)

(a) Summarize in a few sentences the overall *pattern* of tensile and compressive forces in the truss.

(b) What does a structural designer gain by replacing an I-section beam with a truss?

(c) Looking back at Exercise 3.5 and Figure 3.21, how would you expect the forces in the inner links in the truss to vary with the angle they make to the outer links under similar loading conditions?

With reference to the last part of SAQ 3.3, for a given length of truss, a designer is faced with a decision as to whether to use a few inner links at a shallow angle or more links at a steeper angle. Fewer links might be cheaper and weigh less overall. But, as you have seen, they are more highly loaded so they would have to be larger, cancelling out some of the gains in money and weight. Somewhere there is an optimum design of truss but, generally speaking, trusses are considerably over-engineered for the task they have to perform.

If you would like to explore dimensioning a truss to meet a given loading specification, you can do this with the 'Structure design and analysis' package on the T207 CD-ROM.

You have now seen how balancing forces in a static system can provide a simple (if a little repetitive) procedure for estimating the forces throughout quite extensive structural networks. What I want to show you next is how introducing the idea of balanced torques allows a way of finding forces in just part of such a network without having to work exhaustively through each of its components.

3.2 Balanced torques

I am sure that you are familiar with a playground see-saw and with the kind of kitchen scales (for weighing cooking ingredients) usually referred to as a kitchen balance. You might think that the latter is termed a 'balance' because it operates by balancing two weights against each other and similarly for the see-saw. However, this is not strictly the case. The see-saw operates by balancing two torques against each other. Two people on a simple see-saw have to adjust their positions so that the torques (about the fixed pivot or fulcrum) produced by their weights are equal and opposite (Figure 3.23, overleaf).

If the pivot is in the centre of the beam forming the see-saw then the lighter person must be further away from the pivot than the heavier person in order for the torques to balance. If the magnitudes of their weights are W_1 and W_2 and they sit at distances d_1 and d_2 from the pivot, respectively, then the torques balance when

$$W_1 \times d_1 = W_2 \times d_2$$

Rearranging this, I get:

$$W_1 \times d_1 - W_2 \times d_2 = 0$$

ANSWERS TO EXERCISES

Exercise 3.1

(a) The trolley moves away from you in a straight line.

(b) The trolley moves towards you in a straight line.

(c) The trolley rotates anticlockwise (as seen from above).

(d) The trolley rotates clockwise.

Exercise 3.2

It depends where you place your hand. If it is in the centre of the bar, you can simply push forward. If it is to one side, though, you have to push and also 'twist' the bar to resist the tendency of the trolley to rotate away from the side on which you are pushing.

Exercise 3.3

The procedure for combining the three forces as suggested is shown in Figure 3.29. No matter in which order the pairs of forces are taken, the final resultant force will always be the same.

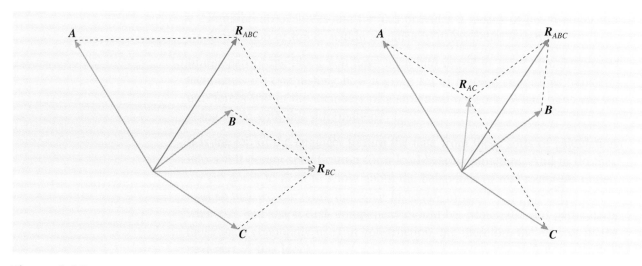

Figure 3.29

Exercise 3.4

(a) The clue to this is to recognize that $\cos\theta = F_x/F$, where F_x and F are magnitudes. The expression we are after is then:

$F_x = F \cos\theta$

Similarly, $\sin\theta = F_y/F$, so:

$F_y = F \sin\theta$

(b) $F_x = F \cos\theta = 300 \text{ kN} \times \cos 25° = 271.9 \text{ kN}$

$F_y = F \sin\theta = 300 \text{ kN} \times \sin 25° = 126.8 \text{ kN}$

Exercise 3.5

The balance of forces in the vertical direction is given by the equation:

$$F_V - T_{1V} - T_{2V} = 0$$

Substituting in the given value for F (20 N) and the new values for the angles of the cords, and recognizing that the tensions in both cords are equal gives:

$$20 \text{ N} - T_1 \cos 45° - T_2 \cos 45° = 20 \text{ N} - T \cos 45° - T \cos 45° = 0$$

which simplifies to:

$$20 \text{ N} - 2T \cos 45° = 0$$

Therefore:

$$T = \frac{20 \text{ N}}{2 \cos 45°} = 14 \text{ N} \quad \text{(to 2 sig. figs)}$$

Exercise 3.6

Balancing the forces vertically I get:

$$F_{CK} \sin \theta + (100 \times \sin \theta) \text{ kN} = 0$$

so

$$F_{CK} = -100 \text{ kN}$$

which is a tensile force acting downwards on joint C.

Adding in the 240 kN compressive force in BC and the 100 kN compressive force in LC, the balance of forces horizontally is:

$$F_{CD} + 240 \text{ kN} + (100 \times \cos \theta) \text{ kN} + (100 \times \cos \theta) \text{ kN} = 0$$

or

$$F_{CD} = -240 \text{ kN} - 60 \text{ kN} - 60 \text{ kN} = -360 \text{ kN}$$

So CD develops a compressive internal force of 360 kN.

Exercise 3.7

The free-body diagram for joint K is shown in Figure 3.30.

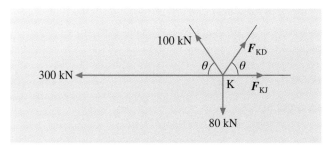

Figure 3.30 Free-body diagram for joint K

Balancing the forces vertically gives:

$$F_{KD} \sin \theta + (100 \times \sin \theta) \text{ kN} - 80 \text{ kN} = 0$$

or

$$F_{KD} \sin \theta + 80 \text{ kN} - 80 \text{ kN} = 0$$

so

$$F_{KD} = 0 \text{ kN}$$

There is no internal force in link KD.

Horizontally:

$$F_{KJ} - (100 \cos \theta) \text{ kN} - 300 \text{ kN} + 0 \text{ kN} = 0$$

so

$$F_{KJ} = 360 \text{ kN}$$

There is a tensile force of 360 kN in this link of the truss.

Exercise 3.8

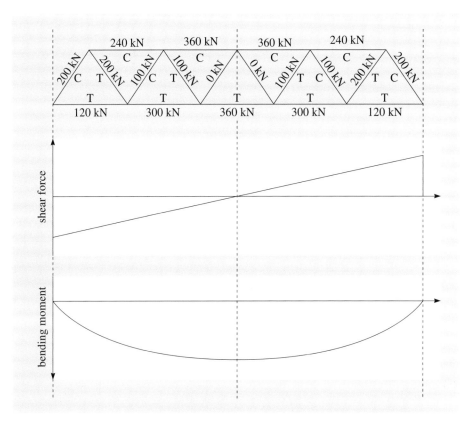

Figure 3.31 Results for the analysis of the 19-member truss, compared to shear force and bending moment diagrams for a beam carrying a uniformly distributed load

Exercise 3.9

Balancing forces vertically:

$$V_1 + V_2 - 30 \text{ kN} - 40 \text{ kN} - 60 \text{ kN} = 0$$

$$V_1 + V_2 = 130 \text{ kN}$$

Balancing torques about the right-hand support (it is equally valid and just as convenient to use the left-hand support) and remembering that anticlockwise moments are positive:

$$-(V_1 \times 18 \text{ m}) + (30 \text{ kN} \times 15 \text{ m}) + (40 \text{ kN} \times 9 \text{ m}) + (60 \text{ kN} \times 3 \text{ m}) = 0$$

which gives:

$V_1 \times 18 \text{ m} = 990 \text{ kN m}$

or

$V_1 = 55 \text{ kN}$

Therefore,

$V_2 = 130 \text{ kN} - 55 \text{ kN} = 75 \text{ kN}$

ANSWERS TO SELF-ASSESSMENT QUESTIONS

SAQ 3.1

As the branches of the rope become more nearly horizontal, the forces in the rope seem to increase. This is similar to the example of the roof truss in SAQ 1.10 where the flatter roof had higher forces in its members than the steeper roof.

My experience tells me that it is extremely difficult to pull the rope perfectly horizontal even when the weight is very small.

SAQ 3.2

If the ropes were horizontal, the 'parallelogram' of forces would have collapsed to a straight horizontal line, thus making the resultant force zero. This could not under any circumstances balance the weight hanging from the rope.

SAQ 3.3

(a) The forces in the outer links in the truss are compressive in the upper surface and tensile in the lower surface. The magnitudes of the forces increase towards the centre of the truss.

The forces in the inner links of the truss alternate between compressive and tensile. The highest forces are in the outer links, with the force decreasing towards the centre of the truss and the centre two links developing no internal force.

(b) A truss should contain less material, overall, than an I-section beam of equivalent performance. Therefore it will be lighter.

(c) Steeper links are likely to experience lower overall forces; shallower links develop higher forces in response to the same loading conditions.

SAQ 3.4

(a) The truss was symmetrical and the applied loads were acting vertically downwards. Thus the reaction forces at the supports were equal, reducing by one the number of values to be calculated. This meant that I could establish the magnitudes of all the external loads (applied loads and reaction forces) simply by balancing the vertical forces.

(b) 'Cut' the truss into two parts so that the forces to be determined are isolated on each side of the cut. Write down one equation each for the equilibrium of vertical and horizontal forces and of torques at a point adjacent to the cut. Calculate the unknown forces from the known values of external loads.

ACKNOWLEDGEMENTS

Grateful acknowledgement is made to the following sources for permission to reproduce material within Part 3 of this block.

Figures

Figure 3.1: Courtesy of View Pictures Ltd.

Course team acknowledgements

Part 3 of this block was prepared for the course team by Joe Rooney and Mark Endean.

Part 4
Materials for structures

CONTENTS

1	Introduction	56
	1.1 Aims	56
	1.2 What's the problem?	56
2	Describing materials	57
	2.1 Property profiles	57
	2.2 Materials and structure	58
3	The material response	61
	3.1 Quantifying the response	62
	3.2 Modelling classes of materials	65
	3.3 Combinations of materials	70
	3.4 Making choices	73
4	Wood – a natural composite	86
	4.1 The nature of wood	86
	4.2 Modelling the properties of wood	90
5	Concrete	92
	5.1 The nature of concrete	92
	5.2 Controlling the properties of concrete	94
6	Steels for structures	101
	6.1 Steels as alloys	102
	6.2 Controlling the properties of steels	105
	6.3 Choosing the right steel	112
	6.4 Specifying steels	113
7	Reinforced concrete	115
	7.1 How reinforcement works	115
	7.2 A design example	119
	7.3 Prestressed concrete	123
8	Summary	125
9	Learning outcomes	126
	Answers to exercises	127
	Answers to self-assessment questions	133
	References	140
	Acknowledgements	141
	Appendix: A world of materials	143

1 INTRODUCTION

1.1 Aims

This part of Block 2 is about the materials used in structures for the purposes of shelter and protection. The overall aims of this part of the block are:

- to illustrate how to begin to make choices between different types of materials on the basis of their mechanical properties;

- to explore the relationships between the nature and composition of certain key structural materials and their behaviour;

- to develop models of the behaviour of materials that can be used to design simple structures.

1.2 What's the problem?

The previous parts of the block have shown various approaches to the building of structures for the purpose of providing shelter and protection. You have seen how external loads create internal forces within structural members and result in stresses which the structure needs to withstand. Various geometrical schemes have been explored for dealing with these stresses in order to maintain structural integrity and prevent the static structure from becoming a dynamic mechanism – not usually a preferred option for the roof over our heads! So, what is the role of materials and what are the material-related problems that the engineer needs to solve?

In Block 1 Part 1 we considered building a tower made from bricks and determined that it could be over two miles tall before the bricks would be crushed under their own weight. This cannot be a realistic height for a real tower, so what considerations are we missing? The result of the model used certainly raises some questions and also hints at the naivety of the model in practical terms. The only limitation considered in our original model was the crushing strength of the bricks themselves.

Exercise 4.1

What other factors, apart from the dead weight of the bricks, should be considered before erecting a tower made from bricks?

The above exercise illustrates the importance of considering the various modes of loading to which a material may be subjected. Adequate models are therefore required, that take account of both the geometry of the structure and the distribution of loads throughout the structure. Factors of safety are also necessary to provide confidence that a structure will remain standing even when subjected to circumstances a little beyond the norm. Section 2 in Part 1 of this block introduced *limit state design*, and the idea that different materials can be attributed different *partial factors* which take account of their predictability in service.

We need to begin to explore the role of materials in providing structural protection by considering how a materials requirement fits into the property profile for a structure, and how this leads through to a design specification. We then need to select appropriate materials based on their responses to applied loads. Then later we shall look in rather more detail at some of the most commonly used materials for structural engineering.

2 DESCRIBING MATERIALS

2.1 Property profiles

Because they are such an integral part of everyday life, we all build up rule-of-thumb knowledge of the different substances that we encounter. For instance, I don't suppose you would have any difficulty distinguishing between wood and metal. But how do you *know* they are different?

The answer is quite simply that different substances *behave* differently when you do things to them. You wouldn't try to fold a sheet of glass as you would a sheet of paper, for example, and you wouldn't normally expect to see car or bicycle tyres made from steel. In what follows I intend to:

- show you how to *describe* different substances more precisely in terms of their physical characteristics and behaviour; and

- demonstrate to you some of the underlying *reasons* why particular substances behave in characteristic ways.

I shall concentrate on substances that are normally used in solid form and whose physical presence is essential for a structure to function properly. These are substances generally referred to in the engineering community as *materials*, and that is how I shall use the term from now on.

Exercise 4.2

To get an idea of the enormous range of materials in common use, spend a few minutes looking at the structures around you, wherever you are reading this text, and see how many different materials you can recognize that are performing some sort of structural role. You need not be specific about what a material is – just note it as being similar to, or different from, the others.

I hope the exercise has demonstrated the large range of structures around us, and the different materials used in their construction. This raises the question: How did the manufacturer in each case settle on one specific material, given such a wide variety to choose from? We can't answer this question for every material, but let's see if we can at least unravel a little of the complexity.

I have already suggested that materials differ from each other in that they behave differently under the same circumstances. Even if one material behaves only slightly differently from another, that's enough to identify it as different. You should be familiar with expressing these differences in terms of a material's *physical properties*. From here we can start to develop a detailed description of the set of properties any material has in order to help us choose the right material for a job or to see why a material was suitable for, or perhaps not quite up to, the job it was supposed to do.

Exercise 4.3

For the materials you identified in Exercise 4.2, write down why they may have been selected for the particular task and note any disadvantages you can see to them being used in that way.

Choosing the most suitable material for a particular job usually comes down to finding a material whose good points outweigh its bad points for the application. This does not just involve thinking about what happens to the structure in use; it also means taking into account the engineering that can be used to arrive at the right shape for the structure in the first place.

This means that it is very unusual to find a single 'best' material for any application. The circumstances and available technologies generally have a strong influence on how you make things, and often the material that is the easiest to process is not the best performer in use and vice versa. Most choices of material represent a compromise between the properties you want and the ones you have to make do with.

We shall be returning to the subject of materials selection later, but first we need to consider how to quantify some of the material properties relevant to structures.

There are close links between the properties of a material and its make-up at an atomic level. If you are not familiar with these relationships and how they provide convenient groupings of materials based on their composition, you should read the Appendix before you go much further in this section.

2.2 Materials and structure

In the previous parts of this block we have considered the requirements of structures that have to bear loads, and explored the ways this can be achieved in terms of configuration and geometry. However, a sound structural design is not enough on its own; we must take care over the choice of materials for building the structure, as illustrated in Figure 4.1.

Figure 4.1 Good structural design and careful materials selection are necessary for protection against the elements

This opens up a wealth of possibilities that are too wide to cover completely here. I shall concentrate, therefore, just on quantifying the requirements for a material and establishing the criteria for its selection. Even if we were only to consider materials that occur naturally in reasonable quantities on Earth, the possibilities and permutations available would be enormous. Combining this with the possibility of using artificial 'engineered materials' can lead to quite a headache for the engineer in the role of a problem solver! Indeed, structural strength and stiffness as discussed in Part 1 may not be the only considerations. Other issues may include resistance to weather, an ability for the materials used to let natural light through, or aesthetic appeal, to name but a few.

At the simplest level the configuration and dimensions of a structure will determine how the external loads are distributed as internal forces within the construction material. The integrity of the structure then comes down to the ability of the materials from which it is assembled to stand up to those forces and the margin for error allowed in the design. But to provide *protection* for the space inside the structure yet more factors may have to be considered. For example, suppose we want to cover an area and enclose a volume above it for the purpose of providing a shelter. Two main issues arise:

- construction of a self-supporting structure;
- covering of the area over the structure.

Figure 4.2 shows two examples which fulfil this specification. Figure 4.2(a) shows the 'cupola' of Florence cathedral, designed by the architect and engineer Filippo Brunelleschi in 1420. This 33 m high dome was constructed around a framework of stone ribs with the spaces between filled by a series of brickwork courses. Each smaller ring of bricks successively tightened the structure and maintained equilibrium while the cupola was being built (a little more stable than the hypothetical brick tower discussed earlier). The structure is estimated to weigh around 37 000 tonnes. In contrast, although of similar geometry but considerably lighter, is the second example shown in Figure 4.2(b). This is the main tower at the National Space Centre in Leicester, designed by Nicholas Grimshaw & Partners and completed in 2001. It consists of a steel skeleton structure covered with a transparent lightweight flexible polymer membrane made of poly(ethylene-tetrafluoroethylene) (ETFE for short) and is 41 m high.

(a) Dome of Florence Cathedral   (b) National Space Centre main tower, Leicester

Figure 4.2 Two different materials approaches to enclosing a space

A clear factor in the construction of these buildings is the availability of materials at the time of construction. ETFE has only been available as a cladding material since the mid-1980s. However, it is still the case that many of today's new buildings are constructed from brick, not so dissimilar to those used for the dome of Florence cathedral. Other buildings such as the

National Space Centre are based on a stiff frame with a flexible covering. This shows that the arrival on the scene of new materials increases the range of options from which the designer can choose rather than heralding the disappearance of any of the traditional ones.

So how do you go about selecting materials when faced with an apparently infinite number of possibilities? This is what we are going to explore in the next few sections.

SAQ 4.1 (Outcome 4.1)

The two designs depicted in Figure 4.2 are both dome-shaped but are structurally quite different. Fill in the gaps in Table 4.1 and then write a few notes about each of the following in the same terms.

(a) an umbrella;

(b) a safety helmet for use on a construction site.

Table 4.1 Two possible types of structure to meet a design specification

Scheme	Brunelleschi's cupola	National Space Centre tower
Structure	Hard shell	Skeleton with flexible skin
Materials properties critical to providing structural integrity		
Additional properties of importance to design		

3 THE MATERIAL RESPONSE

When a material is placed under load, as you know from Part 1 of this block, the load sets up an internal force in the material. Let's look at the three familiar loading conditions and the material response to each. Figure 4.3 shows the responses when a sample of material of the same dimensions (the dotted outline) is placed under pure tension or pure compression. Bending is also shown to illustrate how it results in both tension and compression in the sample. Another important issue is what happens to the materials when the applied load is removed. Do the materials return to their original shapes?

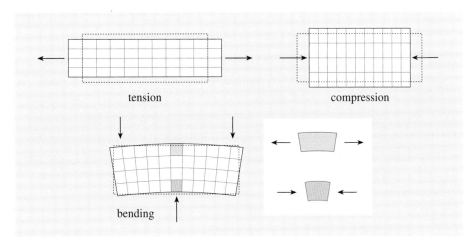

Figure 4.3 Schematics illustrating tension, compression and bending. The inset (lower right) shows how bending can be considered as a combination of tension and compression in different parts of the sample

How a material behaves under each of these loading conditions (and any combination of them) will determine its suitability for use in construction. And a detailed and accurate knowledge of its behaviour is vital for the development of successful designs. In order to assess materials for construction purposes we need methodical tests to establish their behaviour quantitatively.

Exercise 4.4

How do the following objects respond when squeezed between two fingers?

(a) a pencil

(b) an eraser

(c) poster adhesive, such as Blu-Tack®.

Try giving them a quick prod first and then secondly squeezing and holding for several tens of seconds. Is the response different? Consider the material behaviour in terms of how the shape of the object changes when the load is applied, and also whether the shape is permanently changed after the load is removed.

The exercise above has highlighted the responses of the different materials to a load, and the time for which the load is applied. Behaviours can be categorized as follows:

(a) In *elastic behaviour* there is a very rapid recovery to the original, undeformed state on removing the load. The strains are normally small and, as you'll see in the next section, have a linear relationship with stress, so this is also known as *linear elasticity*.

(b) In *viscoelastic behaviour* there is a time-dependent increase in strain with time under load (known as creep). The strains are often larger than in linear elasticity and there is time-dependent recovery towards the original, undeformed state on removing the load (this always takes longer than the period of creep).

(c) In *long-range elastic behaviour* there are very large, but recoverable (elastic), deformations. This is also known as *rubber elasticity* or *high elasticity*.

(d) In *plastic behaviour* the deformation is permanent. Under appropriate conditions, the deformation can be time-dependent. This too is, perhaps confusingly, referred to as creep, although the mechanisms involved are very different from those in viscoelasticity. This type of creep is also known as *viscoplasticity*.

3.1 Quantifying the response

Let's look a little more closely at the requirements for *strength* and *stiffness* as introduced in Block 2 Part 1, Section 3.1. The specifications for most structures will include statements about both of these criteria.

To achieve a structure of the required strength and stiffness a designer can alter both the size (and geometric design) and the materials used. In order to judge the merits of a particular material for the task we need to remove the influences of scale. This is achieved by the use of the concepts that you have already met a number of times in this course – stress and strain. ▼Stress and strain▲ provide two measures which allow us to quantify and compare materials independently of the size and shape of any sample we use for testing.

▼Stress and strain▲

Force and area together are used to define the *stress* in a component. The stress is found by dividing the magnitude of an applied force by the area over which it acts. Thus increasing the cross-sectional area of a component subject to a given force will reduce the stress in the component – a given stress will result from a lower force acting on a smaller area or a larger force acting on a greater area. Two samples of identical material with different cross-sectional areas will fail at the same stress, even though the force required to attain that stress is much higher for the sample with the larger cross-section.

The use of stress therefore allows us to describe the behaviour of the *material* rather than the *component*. How a component responds to a given load will depend on the magnitude of the stress developed within the component.

We use the Greek letter σ (pronounced 'sigma') to represent stress. F is the magnitude of the force, and A is the area over which the force is acting.

Mathematically, we write the definition of stress as:

$$\sigma = \frac{F}{A}$$

The units of stress will be the units of force divided by the units of area; that is, newtons (N) divided by metres squared (m²). The unit is therefore newtons per metre squared, N/m² or N m⁻².

The unit N m⁻² is sometimes called the pascal and given the symbol Pa, but we will not use this terminology.

This brings us to a definition of *strength*. The strength of a material is the maximum stress that the material can withstand before it fails. As stress depends on the area, a smaller sample of material will fail under a lower force but at the same stress as a larger sample. As with stiffness, there are various definitions of strength depending on the manner in which the sample is loaded.

Let's now consider the *strain*. If we apply a tensile force to a material, it will extend in response. This extension is frequently barely perceptible, unless you are pulling something like a rubber band.

Strain is defined as:

$$\text{strain} = \frac{\text{extension}}{\text{original length}}$$

In this way strain also allows us to quantify the material's response to loading, independently of the size of the sample used.

Strain is represented by the Greek letter ε, called epsilon, and the length of the sample by the letter l. Because strain is a ratio of two lengths, it has no units:

$$\varepsilon = \frac{\Delta l}{l}$$

The symbol Δ (Greek capital delta) is used as a shorthand way of saying 'the change in'. So Δl means 'the change in l'. (Δl is said by running the names of the letters together: 'delta el'.

In practice, values of strain are usually quite small, and for this reason they are often expressed as percentages. It's easier to say that 'the strain is 0.1 per cent' than 'the strain is 0.001'. To calculate strain directly as a percentage:

$$\varepsilon \text{ (as a percentage)} = \frac{\Delta l}{l} \times 100\%$$

The greater the initial length of a sample, the more it will extend when subjected to a force. However, whatever the sample's initial length, for a given force the strain will be the same. That is, a particular value of strain in a given material always indicates the same stress.

Exercise 4.5
A panel of an internal ceiling is suspended from four wires each 1 mm in diameter. A modification to the design means that the panel will only be held from three wires. What diameter of wire must be specified to ensure that the stress in each wire is the same as before?

Where there is a linear relationship between stress and strain, the material can be said to obey Hooke's law. This is one of the fundamental laws of elasticity. Hooke actually formulated the relationship in terms of force and extension, and it was Young who modified it to its present form, namely:

$$E = \frac{\sigma}{\varepsilon}$$

As well as Young's modulus E, a stress–strain relationship derived from experiments can be used to define other important quantities and limiting behaviours of materials. Figure 4.4 shows how these are obtained from the results of a standard tensile test in which a specimen is stretched to destruction. In Figure 4.4(a), which is for a steel specimen, the *elastic limit* is the maximum stress in the linear elastic region, and the *tensile strength*, or *ultimate tensile strength* (UTS), is the maximum stress level attained in the stress–strain curve. With one exception, all the stresses in the curve are based on the *original* area of cross-section. The exception is in the case of the *fracture stress*. As there is frequently a significant reduction in area of the cross-section leading up to fracture, fracture stress is calculated from this smaller, final cross-sectional area. So the value of fracture stress is generally higher than is obtained simply by reading a value off the vertical axis of the stress–strain curve.

The stress at the onset of plastic deformation, the *yield stress*, is another important characterizing parameter of materials. Some materials (for example certain types of steel and some plastics) show a well-defined maximum in the stress–strain curve at the onset of plastic deformation. Figure 4.4(b) shows this for a different grade of steel. The stress at which this *yield point* occurs is taken as the yield stress. In other materials, the stress required to produce an arbitrary but defined amount of plastic deformation is taken as an indication of the yield stress. The construction for such a *proof stress* taken at a plastic strain of 0.5% is shown in Figure 4.4(a). Any appropriate value of strain can be used, as long as it is specified; 0.2% strain is another common value. Figure 4.4(b) also shows the *elongation* of the material, the percentage increase in length due to plastic deformation. This is a measure of the *ductility* of a material.

SAQ 4.2 (Outcome 4.2)

Estimate the following from the tensile test results on the steel specimen shown in Figure 4.4(a):

(a) the elastic limit of the steel

(b) the tensile strength of the steel

(c) the 0.5% proof stress of the steel

(d) the fracture stress of the steel if the reduction in area at failure was 10%

(e) the elongation of the steel at failure.

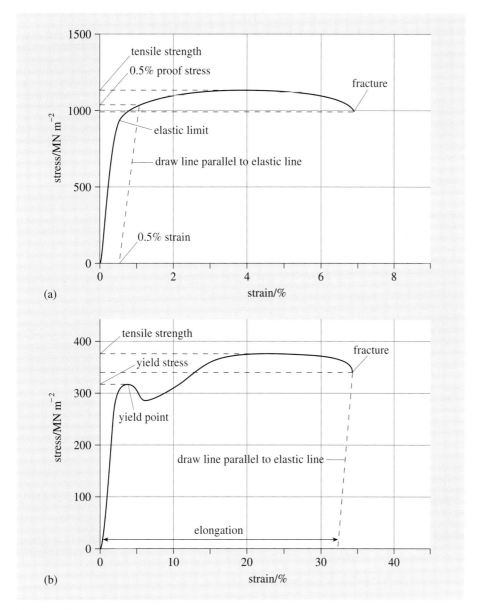

Figure 4.4 Tensile properties from stress–strain curves for two different steels: (a) one without a yield point, and (b) one with a yield point (notice the differences in scales on both axes)

3.2 Modelling classes of materials

Having reminded you of how a material's response to a load can be quantified, I now want to go on to look at how to select suitable materials for a structure. You might immediately think this means identifying the material with the highest value of some critical property, such as tensile strength, but it's not quite that simple, as you will see.

It may seem a bit obvious, but different materials behave differently, as you established for yourself in Exercise 4.4. But it's not just that they have different values of each of the mechanical properties that we have been looking at so far. The *pattern* of their behaviour is also different. Figure 4.5 (overleaf) shows stress–strain curves for four materials which we can compare with the steel specimen discussed in the previous section. But note the differences in scales, both among the four shown here and as compared with the steel data shown in Figure 4.4.

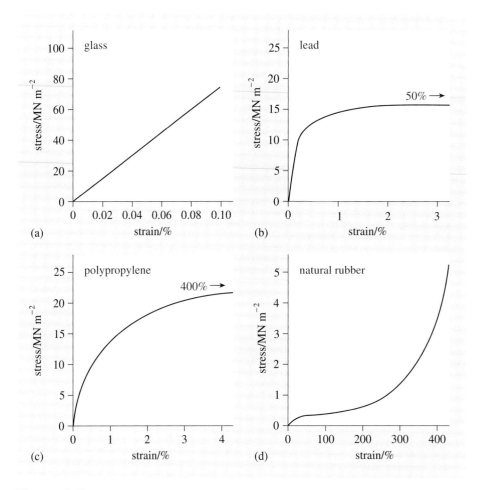

Figure 4.5 Stress–strain curves for (a) glass, (b) lead, (c) polypropylene and (d) natural rubber

The curves show qualitatively the differences in behaviour exhibited by the different classes of material to which they belong. These classes are compared schematically as A to E in Figure 4.6.

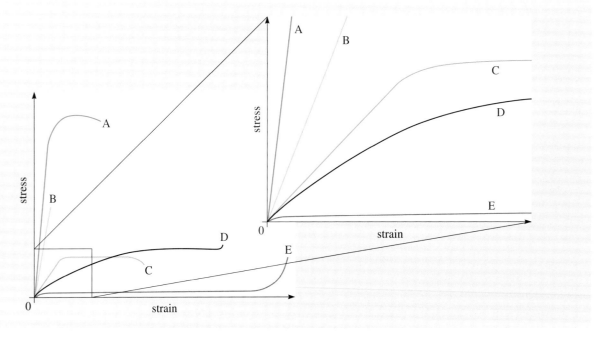

Figure 4.6 Schematic stress–strain curves for different classes of material. The inset (right) shows a close-up of the region close to the origin

SAQ 4.3 (Outcome 4.3)

Identify the curves in Figure 4.6 that best match each of the following materials:

(a) A strong, high-modulus material such as the steel described in the last section, with a high yield stress and some ductility.

(b) A lower-modulus, low yield stress material which is very much more ductile than steel, similar to lead.

(c) A fairly high-modulus material which shows no evidence of ductility or yielding, say similar to glass.

(d) The material with the lowest modulus, similar to natural rubber.

(e) A material with a low modulus and yield stress, that exhibits nonlinear elasticity like polypropylene.

What you can't tell with any certainty from curves such as those in Figures 4.5 and 4.6 is where elastic (recoverable) extension ends and plastic deformation starts. My descriptions were based on foreknowledge.

What further information would you need to distinguish between elastic and plastic deformation?

You would need to see how the materials behaved when they were unloaded from some point before failure.

Tensile property data for the five materials are listed in Table 4.2. I have also included some data for a type of wood commonly used in the construction of modern houses. With the aid of these data, and Figures 4.5 and 4.6, what can you now say about the suitability of these materials for a structural member?

Table 4.2 Typical tensile property data for several materials (approximate values)

Material	Young's modulus $GN\ m^{-2}$	Yield stress/ proof stress $MN\ m^{-2}$	Tensile strength $MN\ m^{-2}$	Elongation %
Steel	210	1050	1150	6.4
Glass	70	—	70	0
Lead	14	11	14	50
Polypropylene	1.5	19	33	400
Natural rubber	0.003	—	50	0
Softwood (parallel to grain)	14	—	50	5

Here 'elongation' implies plastic strain.

SAQ 4.4 (Outcomes 4.2 and 4.3)

Steel is in widespread use in buildings and other structures. In what way would you expect beams made from the other materials mentioned so far (in Table 4.2 and characterized in the various stress–strain curves) to behave compared to steel and how suitable do you think each material would therefore be for this purpose?

So, based on this preliminary analysis, steel emerges as a better material than the others on the whole, although you could easily argue that it's an unfair comparison. But that's still not the entire story by any means. Here are some other points we need to consider:

- Although we noted creep in the lead and polypropylene, and have mentioned fatigue and other time-dependent failure mechanisms (in Part 1 of the block), we haven't examined these in any detail for our range of materials.
- We have not touched upon other, possibly important, properties such as hardness (how difficult it is to scratch or indent the surface), the resistance to wear or abrasion, corrosion resistance and cost.
- The tensile properties of materials in fibre form can be much improved over the bulk properties cited in Table 4.2. This applies particularly to polypropylene and glass. (You might be aware that polypropylene is used in ropes and that glass fibre is used as the reinforcement in composite materials.)
- The structural member's own weight may be an issue, particularly if many are used together, and might be a significant factor in the loading of the building. This is related to the material's density.

This section has concentrated on simple tensile loading of samples of material to quantify a material's response to an applied load. There are other ways of loading, such as shear, that lead to ▼Other moduli▲ but in the next part of this section I want to show you how the stiffness of a structural component can be modified by using different materials in combination.

▼Other moduli▲

A distinction has to be made between mode of loading and the type of stresses induced by the loading. The kind of stresses that are induced by loadings can be broken down into just two types: those which are tending to change only shape and those which are tending to change only volume.

Some modes of loading are conveniently described by elastic moduli different from the one we have so far considered (Young's modulus).

Under shear deformation (Figure 4.7) in the linear elastic region, the *shear modulus*, G, is defined in terms of the shear stress τ (tau) and shear strain γ (gamma) as:

$$G = \frac{\text{shear stress}}{\text{shear strain}} = \frac{F/A}{\gamma} = \frac{\tau}{\gamma}$$

where the shear strain γ, as defined in Figure 4.7, is measured in radians. Note the approximation in Figure 4.7 that $\gamma \approx \tan \gamma$ for small values of γ.

Radian measure is discussed in Block 3 Part 1.

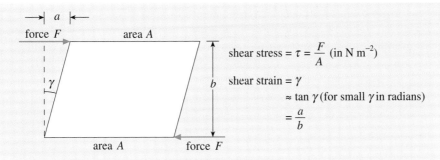

Figure 4.7 Shear stress and shear strain

Under hydrostatic or bulk stresses (that is, those which tend to change only volume so they act equally in all directions, see Figure 4.8) in the linear elastic region the *bulk modulus*, K, is defined as:

$$K = \frac{\text{bulk stress (i.e. pressure)}}{\text{bulk strain}}$$

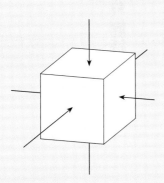

Figure 4.8 Hydrostatic stress

If p is the applied pressure which causes a volume change ΔV in an original volume V, then:

$$K = -\frac{p}{\Delta V / V} = -\frac{pV}{\Delta V}$$

This is negative because a positive pressure causes compressive stresses, which decrease volume. K provides a measure of the *incompressibility* of the material which for solids is high, as reflected in their bulk moduli. Liquids and gases are far more compressible, so K is lower for them.

Exercise 4.6

(a) A 25 mm cube of aluminium (shear modulus $G = 28$ GN m^{-2}) is subject to a shear stress such that one face is displaced by 0.80 mm relative to the opposite face. Calculate the shear force.

(b) If the same cube were subject to a hydrostatic stress, what pressure would be required to produce a volume decrease of 0.01% if the bulk modulus of aluminium is 47 GN m^{-2}?

Based on the data in Exercise 4.6, it's worth noting that, since aluminium has $E = 70$ GN m^{-2}, the bulk modulus is less than the Young's modulus but greater than the shear modulus. This also holds true for most other metals and ceramics but not polymers, natural rubber or even lead. For a given material the three elastic moduli are related. You might expect them to be, since they describe the elastic behaviour of the same substance.

3.3 Combinations of materials

It is increasingly common to find materials being used in combination with each other to obtain the benefits of the good properties of each constituent (see 'Composite materials' in the Appendix to this block). One characteristic of composite materials is that they can be highly anisotropic – their properties can vary dramatically depending on the direction of application of a load, for example. This arises because the way that the properties of each material combine to provide an overall property for the composite depends on how each material is dispersed with respect to the others. I am going to look here at how it is possible to estimate the Young's modulus of a composite material, taking these factors into account.

In order to derive an expression for the modulus of a two-component composite material in terms of the moduli of the individual components, the composite can be treated as two blocks of material in intimate contact (Figure 4.9). The volume of each block represents the volume of that component in the composite lumped together.

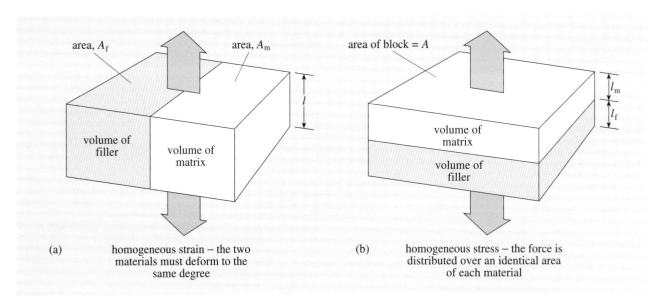

Figure 4.9 A simple model of a composite material showing the two different loading cases, (a) and (b)

This simplified composite block can be loaded in one of two ways. In the first case, (a), the effect of the applied force will be to induce equal strain in each component, since the blocks are firmly fixed together and they are of equal length. This is referred to as the *homogeneous strain case* ('homogeneous' simply meaning 'the same'). You may recognize that this arrangement loads the material components 'parallel' to each other, which leads to an alternative description of this as the *parallel case*. It is also called the Voigt model of loading, for historical reasons.

In the second case, (b), the force is distributed across an identical area for both blocks so the stress in each is the same – the *homogeneous stress case*. The blocks are now loaded 'in series' with each other, so this is also known as the *series case* and is called the Reuss model. I shall treat each of these cases in turn. The notation I shall be using is as follows:

- subscripts m, f and c distinguish between matrix, filler and composite, respectively;
- *l* is the length of the side of the block, parallel to the applied load;
- *A* is the area of the block, perpendicular to the applied load;
- the volume of each block is given by *V*;
- *F, E, σ* and *ε* are force, modulus, stress and strain respectively.

Homogeneous strain

Our starting point in parallel loading (Figure 4.9a) is that each component will be deformed by a similar amount, since they are firmly fixed to each other. The lengths of the blocks are the same in the direction parallel to the applied load, which in effect means they exhibit the same *relative* deformation, which is strain. So:

$$\varepsilon_c = \varepsilon_f = \varepsilon_m = \varepsilon$$

We want to know the modulus of the composite, and this will be:

$$E_c = \frac{\sigma_c}{\varepsilon_c} \qquad (4.1)$$

The overall load applied will be shared between the matrix and filler, so it is possible to say that:

$$F_c = F_m + F_f \qquad (4.2)$$

The stress on each component is the force divided by the cross-sectional area, $\sigma = F/A$, which can be rearranged to give the force in terms of the stress and area, $F = \sigma A$. Substituting this into Equation (4.2) gives:

$$\sigma_c A_c = \sigma_m A_m + \sigma_f A_f \qquad (4.3)$$

Using Equation (4.1) to provide an expression for the stress in terms of modulus and strain, $\sigma = E\varepsilon$, and substituting this into Equation (4.3), gives:

$$E_c \varepsilon_c A_c = E_m \varepsilon_m A_m + E_f \varepsilon_f A_f \qquad (4.4)$$

Since in this case the strains are all equal, this whole equation can be divided through by *ε* to give:

$$E_c A_c = E_m A_m + E_f A_f \qquad (4.5)$$

Now the volume of each component will be its area multiplied by its length, $V = Al$, and that means that $A = V/l$. Equation (4.5) can then be written as:

$$\frac{E_c V_c}{l_c} = \frac{E_m V_m}{l_m} + \frac{E_f V_f}{l_f}$$

The lengths are all equal, $l_c = l_m = l_f = l$, so we can multiply all the terms in this expression by *l* to give:

$$E_c V_c = E_m V_m + E_f V_f$$

Finally, we divide each term by V_c so that the volume of each component is expressed in terms of the total volume of the composite – what is referred to as the *volume fraction* of that component and given the symbol v. The final equation is then:

$$E_c = E_m v_m + E_f v_f \tag{4.6}$$

You may also see this equation referred to as the 'law of mixtures'.

Homogeneous stress

In the second case the blocks are loaded in series (Figure 4.9b). The area of each block perpendicular to the applied load is identical, and they are each subject to the same force, so the stress in each block will be the same:

$$\sigma_c = \sigma_m = \sigma_f = \sigma$$

The extent to which the block deforms under the applied load will be the sum of the individual deformations of each component:

$$\delta_c = \delta_m + \delta_f$$

The absolute deformation of the blocks is related to the strain in each block by multiplying by its length to give:

$$\varepsilon_c l_c = \varepsilon_m l_m + \varepsilon_f l_f \tag{4.7}$$

The strain can be expressed in terms of modulus and stress, $\varepsilon = \sigma/E$, so this can be substituted into each component of Equation (4.7) to give:

$$\frac{\sigma_c l_c}{E_c} = \frac{\sigma_m l_m}{E_m} + \frac{\sigma_f l_f}{E_f}$$

Since the stresses are all equal in this case ($\sigma_c = \sigma_m = \sigma_f = \sigma$), each term can be divided through by σ giving:

$$\frac{l_c}{E_c} = \frac{l_m}{E_m} + \frac{l_f}{E_f}$$

As in the previous case, the goal is to relate the modulus of the composite to the volume fraction of each component, so the next step is to replace l in each term by V/A. That gives:

$$\frac{V_c}{E_c A_c} = \frac{V_m}{E_m A_m} + \frac{V_f}{E_f A_f}$$

We know the areas of each block to be the same ($A_c = A_m = A_f = A$), so we multiply through by A leaving:

$$\frac{V_c}{E_c} = \frac{V_m}{E_m} + \frac{V_f}{E_f}$$

Finally, as before, the volume of each block is converted to the volume fraction by dividing each term by V_c to give:

$$\frac{1}{E_c} = \frac{v_m}{E_m} + \frac{v_f}{E_f} \tag{4.8}$$

Equations (4.6) and (4.8) for combining the properties of components in series or parallel depending on the circumstances and environment find analogues in other branches of engineering that you may come across.

Exercise 4.7

A composite material consists of a matrix with a Young's modulus of 30 GN m^{-2} and a filler material with a Young's modulus of 100 GN m^{-2}. Consider the two loading cases, using the 'series' and 'parallel' models. Sketch graphs of the Young's modulus (vertical axis) over a range of volume fractions of filler (horizontal axis) from 0 (the matrix alone) to 1.0 (the filler alone). Use the same axes for the two cases. (*Hint:* You need only use Equations (4.6) and (4.8) to calculate values of the modulus at a few intermediate values, say 0.3, 0.5 and 0.8, to establish the general nature of the relationships.)

Write a couple of sentences summarizing what you have discovered from this exercise about the effectiveness of 'fillers' in composite materials.

SAQ 4.5 (Outcome 4.4)

How would you have to arrange the filler material in a composite to ensure that it behaved as closely as possible to the homogeneous strain, parallel model?

All of this gives us a vocabulary for the short-term mechanical behaviour of materials and allows numbers to be attached to the various measures of performance such as modulus and tensile strength. You have also had the opportunity to think about comparing materials with each other. What I am going to do next is to look much more closely at how to make precise comparisons between materials, with the aim of selecting the right material for a given task.

3.4 Making choices

I have contrasted the behaviour of a few different materials in terms of their stress–strain curves and moduli. This provides a means of assessing the suitability of materials for use in structural components. I discussed different materials requirements for domes earlier in the block. One design solution relied on a stiff, hard, shell structure while the other relied on a stiff frame covered in a flexible material. The aim of this section is to find a way of navigating around and selecting from the wealth of materials around us.

SAQ 4.6 (Outcome 4.1)

Identify the major structural components and important materials requirements for the following:

(a) a garden shed

(b) a greenhouse.

Nature provides us with many candidate materials which have been used in construction throughout history. If you travel around the area where you live, and further afield, you will see that the local traditions of building were influenced heavily in the past by the availability of materials in the area. Where I live, the older buildings are mostly constructed in brick. Further east, timber framing becomes much more common; in the north and west, the dominant material is stone.

In more recent times the increasing ability to engineer materials with enhanced structural properties leaves us with a bewildering wealth of materials choices. More modern buildings share few local characteristics – except where the planners and architects have made attempts to fit in with the surrounding environment, generally by the cosmetic use of cladding rather than in the structural design.

In selecting materials for a structure we need to make choices based on several criteria which are often in opposition to each other. For example, properties such as modulus, strength and density will be important for structural design. A straightforward design solution to make something stiff may involve using a very dense material. The resulting weight may conflict with the overall design specification for the structure. Also, issues such as material cost and manufacturing cost cannot be ignored and may well dominate the fine-tuning of materials choices.

A useful construct for identifying candidate materials where two different characteristics of the material are important is a materials selection chart developed by Ashby at Cambridge University in the UK. The two materials properties are cross-correlated and plotted onto a two-dimensional graph that is now commonly referred to as an *Ashby chart* (Ashby 1999, Benham *et al.* 1996). Figure 4.10(a) is an example of an Ashby chart, in this case showing Young's modulus plotted against density for several classes of materials. Note that the scales on both axes of the graph are logarithmic; therefore materials which may appear close on the chart can have quite different values of that property.

Let's suppose we are beginning the design of a structural member. The initial specification states that the member needs to deflect as little as possible under applied loads in order to maintain structural integrity. A class of material with a high Young's modulus would indicate a stiff material, so we would be looking towards the upper part of the chart in Figure 4.10(a) for a suitable candidate.

If modulus were the only selection criterion, the material with the highest modulus would always be used. But you know that's not the case in most situations, judging by the wide range of design solutions to most problems. So how do we add a second material property into the equation?

Let's use the mass of the member as we did for bicycle frames in Block 1. The mass of a structural member determines its cost once a material has been chosen, but it may also be important if the overall weight of the structure is of concern in the design of the load-bearing parts.

The mass per unit volume of a material is its density, so a lower-density material might be preferred as long as the modulus criterion can be satisfied. We therefore use density as the second axis on the chart.

The location of a material on the Ashby chart now gives an impression of the relative performance of each material against a combination of the two selection criteria – low-density materials with high modulus will appear towards the top left of the chart, high-density materials with high modulus towards the top right, and so on. This is mapped schematically in Figure 4.10(b). It is important to recognize when looking at Ashby charts that materials with *equally good* performance with respect to the pair of properties being mapped will lie on a line with a particular gradient depending on the situation being modelled. I shall come back to this shortly.

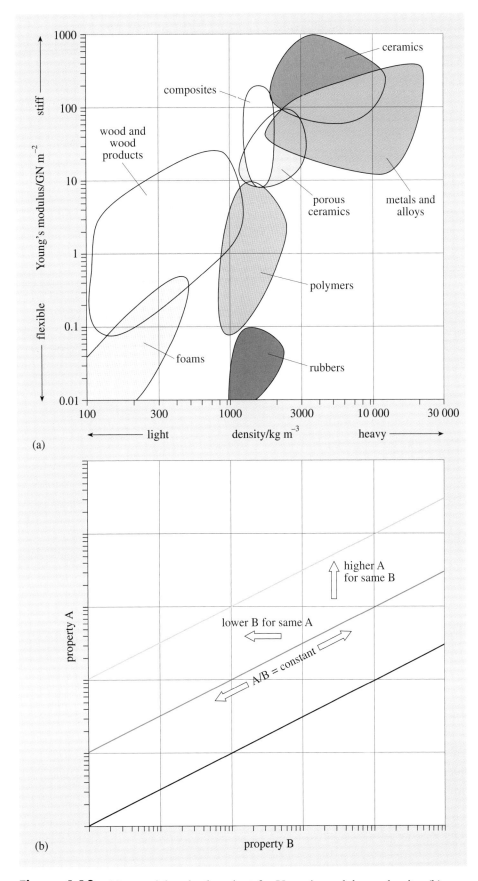

Figure 4.10 (a) materials selection chart for Young's modulus v. density, (b) schematic two-property selection matrix for two properties, A and B. Logarithmic scales are used on both axes

In our search for a low-density/high-modulus material, therefore, we are guided towards the top left of the chart. Looking again at the plot of materials shows that wood does well on the density front at lower modulus. Building materials such as brick and concrete will lie within Ashby's 'porous ceramics' category, which perform comparably to wood but at higher moduli and higher density. Composites, though, outperform both groups. To help you visualize this you might like to draw the relevant gradient on the graph in Figure 4.10(a) by joining the points (100, 1), (1000, 10) and (10 000, 100).

Let's look next at a different pair of properties: strength and cost. Figure 4.11 shows these properties on a materials selection chart for the same classes of materials as in the previous chart. Remember these are only schematic diagrams – strength may be either yield strength or tensile strength.

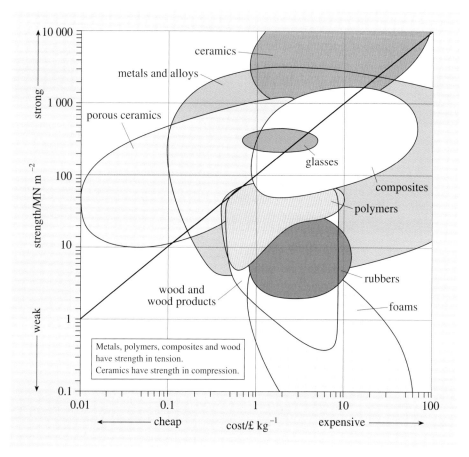

Figure 4.11 Materials selection chart for strength v. cost

Because of the scales used, if there is a linear relationship between strength and cost the relevant diagonal line lies closer to 45° on this chart. Wood now shows up rather poorly and porous ceramics do extremely well, albeit at low strengths. That's because they are so cheap! Composites now show up as better than wood, but expensive. What stands out for me, though, is how metals have appeared on the scene as providing good properties at moderate cost.

However, as is often the case, constructs such as a materials selection chart don't always tell the whole story. The high strength of ceramics is attributed to their ability to withstand compressive loading. In tension (which also, of course, is important in bending) they are very brittle materials and often fail under a very modest load. This is at least a factor of ten less than in compression, owing to microscopic cracks or other imperfections in the material (see ▼Cracks, holes and toughness▲). So, for a structural member which might be loaded in tension or bending, the ceramics would not be the preferred choice.

▼Cracks, holes and toughness▲

If a stress develops in a material as a result of an applied load, there is a critical value of the stress at which a crack of any particular length will start to grow. An alternative view is that there is a critical length of crack that will start to grow at any particular stress.

A widely used and successful approach to the study of the fracture of materials relates the stress (σ) at which a crack of length a will start to propagate in a material of Young's modulus E to the energy used to enlarge the crack. We can quantify *toughness* as the energy used per unit area of the newly created crack surface, and give it the symbol G_C; it has units of joules per square metre, $J\,m^{-2}$. Toughness is specific to the material and is related to the atomic-level structure of the material. The equation relating these parameters turns out to be:

$$\sigma = \sqrt{\frac{EG_C}{\pi a}} \qquad (4.9)$$

which is to say that the stress at which a particular crack will start to grow is inversely proportional to the square root of the length of the crack.

This relationship will be valid as long as the majority of the material remains elastic rather than plastic. Equation (4.9) can be rearranged in terms of the crack length. This shows that there is a critical crack length a_c for any value of stress for a given material:

$$a_c = \frac{EG_C}{\pi \sigma^2} \qquad (4.10)$$

From this relationship you can see that as the stress in a material increases, the length of crack that can be tolerated without the crack growing decreases in proportion to the square of the stress.

Note that the terms describing the material properties in Equation (4.10) are often combined as the so-called *fracture toughness* K_c where:

$$K_C = \sqrt{EG_C} = \sigma\sqrt{\pi a_c} \qquad (4.11)$$

It is this toughness that is quoted in most textbooks and data sources as it is a more useful parameter than G_C when evaluating the effect of cracks on the strength of stressed components.

Young's modulus and toughness values for a range of materials commonly used in construction are given in Table 4.3, listed in order of decreasing toughness.

Table 4.3 Typical toughness values for a range of materials

Material	Young's modulus E/GN m^{-2}	Toughness G_C/kJ m^{-2}	Fracture toughness K_C/MN m$^{-3/2}$
Mild steel	210	100	145
Aluminium alloys	70	20	38
Woods (across the grain)	0.2–1.7	8–20	1.3–5.8
Polymer composite (polyester/glass-fibre)	20	10	14
Woods (along the grain)	6–16	0.5–2.0	1.7–5.7
Concrete	16	0.03	0.7
Window glass	72	0.01	0.9

Equations (4.9) to (4.11) strictly apply only to infinite bodies subjected to uniform stress fields. This is a reasonable approximation if the crack is a lot smaller than the body, but most practical structures also contain features which tend to concentrate the effect of imposed stresses. It is very common to encounter substantial flaws in the form of voids, especially in materials such as concrete cast on-site, and these have the effect of disrupting the uniform distribution of stress. Such flaws significantly affect the local stresses that act on any cracks present. Figure 4.12 shows an elliptical hole in a plate subjected to a force at right-angles to the hole's major axis. The length of the major axis is $2c$ and the minimum radius of curvature of the hole is r.

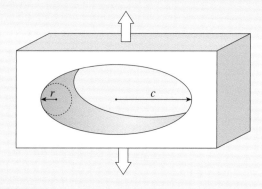

Figure 4.12 An elliptical hole in a rectangular plate

Above and below the hole, the load will be spread over the whole area of the plate, but adjacent to the hole the area is greatly reduced, leading to an increase in the stress in the material. It can be shown that the maximum stress in the material σ_m is related to the average stress in the plate σ_a by:

$$\sigma_m = \sigma_a \left(1 + 2\sqrt{\frac{c}{r}} \right) \qquad (4.12)$$

where the expression in brackets is known as the *stress concentration factor*.

So for a circular hole (when $c = r$) σ_m is simply $3\sigma_a$. For all large values of c/r (flatter, broader defects) the stress concentration factor approximates to $2\sqrt{(c/r)}$, and as r decreases still further relative to c (in other words the hole is getting closer to being a sharp crack) the maximum stress at the edges of the hole approaches infinity.

Whether or not a defect does cause a problem in practice depends very much on the characteristics of the materials involved. Consider, for example, the effect of a flaw with circular cross-section, in a joint under load (Figure 4.13)

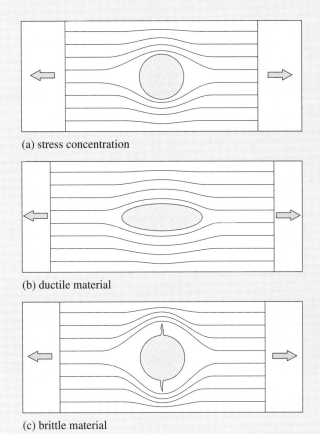

(a) stress concentration

(b) ductile material

(c) brittle material

Figure 4.13 The effect of a flaw in ductile and brittle materials

If you imagine the lines in the figure to suggest the transmission of the applied load through the material, the spacing of the lines gives an indication of the stress in the material. A circular flaw has the effect of concentrating the stress, as suggested by the closer spacing of the lines in Figure 4.13(a). If the stress in the material exceeds its yield stress at this point, plastic deformation will occur, increasing the radius of curvature of the flaw at the critical point and reducing the stress concentration until the stress drops below the yield stress as in Figure 4.13(b).

If the material is not ductile the stress can exceed the failure stress of the material. A crack then develops at the point of highest stress, further concentrating the stress and leading to rapid failure, as in Figure 4.13(c).

Even in the ductile material there is a significant chance that the stress concentration caused by such a flaw may coincide with a pre-existing crack. If this has the right dimensions it will start to grow before the material either yields or cracks at the flaw itself, and the result is catastrophic failure. You will study more about these kinds of situation in Block 5.

> **SAQ 4.7 (Outcome 4.3)**
>
> Contrast the suitability of a ductile metal versus a ceramic as the material of choice for: (a) a beam (b) an arch.

The conclusions reached in my answer to SAQ 4.7 may not have come as much of a surprise to you. We know from our everyday experience that materials like metals are usually quite stiff and strong. But so far we have only narrowed the selection down to one or two classes of materials. We have not yet fully exploited the usefulness of the charts. Let's just spend a moment linking the choice of material rather more precisely to the way it will be used – and specifically, how it will be loaded.

Figure 4.14 shows a chart of Young's modulus against density, but with a rather more detailed breakdown of material types. The graph includes a number of diagonal reference lines, some with a gradient of 1 (on log scales) but others with gradients of 2 and 3. The purpose of these lines and their usefulness begins to become apparent when you look more carefully at the ▼Merit indices▲ that you first encountered in Block 1.

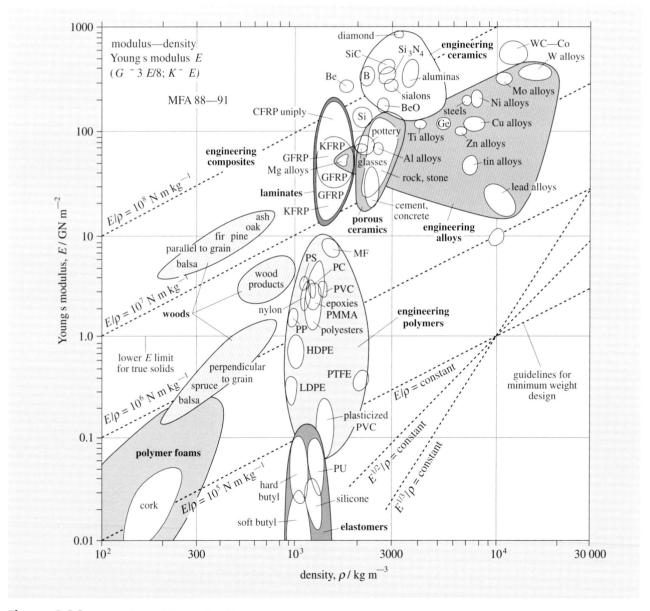

Figure 4.14 Young's modulus v. density

▼Merit indices▲

One way of making comparisons between materials based on a combination of two of their physical properties is to combine the properties together into what is termed a *merit index*. Materials that perform equally well will have identical values of merit index, and the values obtained for different materials will enable them to be ranked in order of suitability.

The way in which the properties of interest combine to provide a merit index will depend on the way in which the material is to be used. Merit indices are therefore specific to particular applications – in our context that predominantly means the way in which load is applied to a structural component.

Take, for example, a simple tie, loaded in pure tension. To find how to choose the material that will give the tie of lowest weight to provide a specified extension under a given load, we can work out how the Young's modulus *scales* with the density of the material.

When working with merit indices, it is important to establish what aspects of the design must be fixed in order for a comparison to be made. For this example, I am going to use a tie of fixed length L subject to an applied load F and permitted a maximum extension e. The cross-sectional area of the tie, A, will vary depending on the amount of each material required to provide the stated stiffness. This clearly depends on the Young's modulus E and gives us a 'handle' on the weight of the tie via its density ρ.

The following procedure is a good way of deriving a merit index:

Step 1: Write down any true expression for the attribute being sought.

Step 2: Group together those parameters that are either materials properties or fixed quantities (in other words, separate the constants from the variables).

Step 3: The merit index is then just the combination of properties that remains.

This is how it works for the tie.

Step 1: Write down any true expression for the attribute being sought.

We are seeking the tie of lowest weight mg. Mass is density times volume, and the volume is derived from the dimensions of the tie, giving us:

$$mg = \rho g A L$$

Merit indices work by maximizing the attribute being sought which, in this case, is lightness rather than weight. So we need to use the reciprocal of weight:

$$\frac{1}{mg} = \frac{1}{\rho g A L} \tag{4.13}$$

We need to bring in the other materials property of interest, Young's modulus. That's related to area through Hooke's law (another way of saying $E = \sigma/\varepsilon$):

$$E = \frac{F}{A} \times \frac{L}{e}$$

Rearranging to make A the subject of the equation:

$$A = \frac{FL}{Ee}$$

and then substituting back into Equation (4.13) gives:

$$\frac{1}{mg} = \frac{Ee}{\rho g FL^2}$$

Since g appears on both sides of the equation, these cancel out to leave:

$$\frac{1}{m} = \frac{Ee}{\rho FL^2}$$

Step 2: Separate the constants from the variables.

The extension, applied force and length are all fixed, so these can be brought together into a single expression:

$$\frac{1}{m} = \frac{e}{FL^2} \times \frac{E}{\rho}$$

Step 3: The merit index is the combination of properties that remain.

So the merit index tells us that the material that will provide the lightest tie for a given set of conditions is the one with the highest value for E/ρ.

In bending, the situation is a bit more complicated. For a start, if both the breadth and depth of the beam can be altered to fit the design brief, an almost infinite number of solutions exists. One of those dimensions must be fixed to make the calculations manageable. Then if you look back to Table 1.2 in the first part of this block you will see that the second moment of area, and hence the stiffness, of a rectangular beam depends on both its width and depth. However, the scale of dependence is different – the second moment of area depends on the *cube* of the depth but only *linearly* on the breadth.

Let's go through the motions of deriving a merit index again, but this time for a rectangular beam of fixed length L, fixed width a, and variable depth b. The beam is to have a defined bending stiffness, so for a given bending moment M the radius of curvature, R, of the beam will be constant. We are trying to find the lightest beam that will meet the design specification.

Step 1: Write down any true expression for the attribute being sought.

As before, we are designing to minimum weight and the reciprocal weight is the reciprocal of the material density times its volume and g. For the beam this is given by:

$$\frac{1}{mg} = \frac{1}{\rho gabL} \qquad (4.14)$$

The Engineer's Bending Equation (in Part 1) provides a way of relating the dimensions of the beam to the Young's modulus of the material:

$$\frac{M}{I} = \frac{E}{R}$$

or

$$I = \frac{MR}{E}$$

From Table 1.2 you can see that $I = ab^3/12$, so that gives:

$$ab^3 = \frac{12MR}{E}$$

We need to get this into a form where we can substitute something back into Equation (4.14) to introduce E into the equation. That's most easily done by multiplying both sides by a^2, since this is one of the fixed parameters:

$$a^3b^3 = \frac{12MRa^2}{E}$$

which can be reduced to:

$$ab = \sqrt[3]{\frac{12MRa^2}{E}}$$

This is a bit unwieldy, and it is better to use the alternative notation:

$$ab = \left(\frac{12MRa^2}{E}\right)^{1/3}$$

Substituting this expression for ab into Equation (4.14) gives:

$$\frac{1}{mg} = \frac{1}{\rho gL}\left(\frac{E}{12MRa^2}\right)^{1/3}$$

Once again, g can be eliminated from both sides to leave:

$$\frac{1}{m} = \frac{1}{\rho L}\left(\frac{E}{12MRa^2}\right)^{1/3}$$

Step 2: Separate the constants from the variables.

The two variables in this equation are E and ρ, as planned. Bringing them together leads to:

$$\frac{1}{m} = \frac{E^{1/3}}{\rho}\left(\frac{1}{L\left(12MRa^2\right)^{1/3}}\right)$$

Step 3: The merit index is the combination of properties that remain.

So now we have a new merit index that tells that the material that will provide the lightest beam of a fixed width is that with the highest value of

$$\frac{E^{1/3}}{\rho}$$

Now try the following exercise to see the difference when the depth of the beam, rather than its width, is predetermined – a much more realistic specification in most practical circumstances.

Exercise 4.8
Derive a merit index for materials to meet a given bending stiffness requirement for a simple rectangular beam whilst minimizing the weight of the beam. The length (L) and depth (b) of the beam are fixed. The width (a) here is variable.

Using a merit index approach, you can see that two materials with identical values for the merit index that describes a particular set of circumstances will be equally effective for that application. Better materials will have higher values for the merit index, and materials with a lower merit index will be less desirable. So if we can find a way of applying the merit index to the Ashby chart, we will be able to see at a glance how the materials on the chart compare.

If you look back to Figure 4.14, you can see three diagonal lines passing through the point $\rho = 10^4$ kg m^{-3}, $E = 1$ GN m^{-2}. The shallowest line joins all the points on the graph with the same value of E/ρ as this point. The line is effectively a line graph of

$$\frac{E}{\rho} = \text{constant}$$

The line you are looking at has a particular value of E/ρ, which is 10^5 N m kg^{-1} in this instance. Other values for this merit index will then have exactly the same *gradient* as this line when plotted on these log–log axes but pass through different points on the graph. Some lines with higher values of E/ρ have been included on the graph for reference.

The steepest line that passes through the point $\rho = 10^4$ kg m^{-3}, $E = 1$ GN m^{-2} joins points with the same value of $E^{1/3}/\rho$. Just as with the other merit index, lines drawn parallel to this one will join points which share similar values for $E^{1/3}/\rho$.

The third line is for a merit index of $E^{1/2}/\rho$. This can be used in two circumstances:

- when designing lightweight beams against bending where neither the width nor the depth of the beam are fixed but the ratio of width to depth is constant;
- when designing lightweight struts against buckling.

You can now use these lines to make quick comparisons between different materials. Let's use them to examine the two different beam design 'envelopes' that you looked at in 'Merit indices'.

Exercise 4.9

I want you to find materials that perform as well as, or better than, steel in the design of a rectangular cross-section beam. The length of the beam is predetermined. As before I want you to explore two different design constraints:

Case 1: the depth of the beam is fixed but its width can be varied;

Case 2: the width is fixed but the depth can be varied.

By choosing an appropriate merit index for each case and drawing lines of the respective slopes through the 'steels' area on the Ashby chart in Figure 4.14, find two materials that would perform as well as, or better than, steel in each instance.

SAQ 4.8 (Outcome 4.5)

Summarize what you have discovered about designing beams from Exercise 4.9. Do you find the outcome convincing? Which aspect of the design calculation is unrealistic?

This further exploration of merit indices has given you a little insight into quantitative approaches to selecting materials for structures. I am sure you can see how different combinations of properties can be used to describe different design problems and how these can be developed into merit indices and combined with appropriate materials selection charts.

Earlier you saw ways of describing materials through property profiles and how materials can be specified to fulfil a particular structural function. You have also seen how these approaches are often simplistic and the actual constraints on real designs mean that judgement must be exercised and compromises made.

In the following sections we shall delve a little more deeply into the world of structural materials that have stood the test of time: woods, concrete and steels (alone and as reinforcement for concrete).

4 WOOD – A NATURAL COMPOSITE

Wood, or 'timber', is one of our oldest structural materials. Its use in buildings stretches far back into antiquity, and it is still a key structural material in many of today's buildings. The mass of wood processed and incorporated into products (mostly associated with the construction industry) is similar to that of iron and steel, which, in terms of volume, means about ten times greater (and that's excluding its use as fuel).

Being a natural material, wood's properties cannot be manipulated to anything like the extent of those of many other materials. Since there are some 30 000 species of tree, each with its own property profile, properties can be selected over quite a wide range through choice of species. But, within any given species, there is a large, and to some extent unpredictable, variability. Design with timber, and in our context, load-bearing design, has to take this variability into account.

4.1 The nature of wood

There's little in the external appearance of a tree to suggest that wood is not only a composite material, but also highly anisotropic – that is to say, having properties that depend on direction. Consider how much easier it is to split a piece of wood along its grain compared to across the grain. Essentially, wood is a foam with elongated cells (the 'bubbles' in the foam) that behave differently depending on the direction in which they are loaded. The walls of the cells are themselves a composite with anisotropic behaviour, as you will see shortly. Wood is also used to make a range of ▼Composite timber products▲ as well as paper. The cell wall structure is based on the material cellulose, the most abundant organic polymer on Earth, that is also a major constituent of natural fibres such as flax and cotton.

▼Composite timber products▲

As versatile and useful a structural material as it is, wood nevertheless suffers from some deficiencies. Its variability is one, and its propensity to structural flaws such as knots is another. Its very high anisotropy can be a problem too, not only in its mechanical properties, but also because moisture-induced swelling and shrinkage are anisotropic, resulting in differential changes in dimensions that lead to warping. Finally, its maximum width is limited by the diameter of the tree from which it is cut. Several composite wood materials have been developed which help alleviate some of wood's deficiencies. Most of these are in the form of sheet or board.

Plywood was known to the Egyptians, but wasn't exploited on a large scale until the 1930s. In this, thin layers (veneers or plies) of wood are laminated together with an adhesive so that the grain direction runs at right-angles in successive layers (Figure 4.15a). Plywoods always have an odd number of plies, so that they are balanced about the mid-plane of the central layer. Increasing the number of plies reduces the anisotropy, but increases the cost, so 3-ply and 5-ply are most commonly used.

PART 4 MATERIALS FOR STRUCTURES

Figure 4.15 Timber composites and their microstructures: (a) plywood, (b) chipboard, (c) fibreboard, (d) MDF, (e) laminated timber portal frames supporting the roof over a swimming pool. (f) and (g) show close-up views of the laminated timber

Chipboard (Figure 4.15b) was developed to make use of waste timber. Resin-coated chips of wood are moulded between flat, heated plates to produce sheets of particulate composite. Chipboard's properties, although isotropic, are markedly inferior to those of plywood. But it is very much cheaper since it can be made from almost any scrap wood.

Fibreboard (Figure 4.15c) is another inexpensive wood composite that starts with wood chips. About 25% of the volume of wood is made up of a resinous substance called lignin, which is thermoplastic. This is softened by pressurized steam so that the chips can be separated into fibres. The chips are formed into a mat which, when hot-pressed, becomes a board bonded by the lignin.

Medium density fibreboard (MDF, Figure 4.15d) is now very widely used in the manufacture of many items of furniture. It is typically made as a board and has the advantage of high strength, drawing on the properties of the wood fibres. Fibres are bound together with a resin under high pressure and a finely textured surface is achieved. Because of its fine texture and isotropy it is frequently processed into a broad range of profiles for structural and decorative purposes.

Prefabricated laminated timber (Figure 4.15e) is used for such products as shaped beams in the construction industry.

Table 4.4 lists some typical properties of the above materials compared with one of the softwoods (Douglas fir). The symbols denote the properties parallel ($\parallel$) and perpendicular ($\perp$) to the grain, where applicable.

Table 4.4 Typical properties of some wood composites and natural woods.

	Density kg m^{-3}	Flexural strength GN m^{-2}			Young's modulus GN m^{-2}		
	ρ	$\sigma(\parallel)$	$\sigma(\perp)$	σ	$E(\parallel)$	$E(\perp)$	E
Douglas fir	590	54	3.2		10	1.1	
Plywood							
3-ply	520	73	16		12	0.9	
9-ply	600	60	33		11	3.3	
Chipboard	680			18			2.8
Fibreboard	1000			43			3.5
MDF	680			40			3.5

Source: Dinwoodie, J.M. (1981) *Timber, its Nature and Behaviour*, Van Nostrand Reinhold

Wood is made up of a very large number of cells, the building blocks of all living organisms. It has a complex structure at different levels of scale. To explore this, let's look at a typical softwood, since the structure is simpler than that of a hardwood. Figure 4.16 shows some of the main structural features apparent at scales between a complete tree (approximately 50 m high) and the unit in crystalline cellulose (approximately 1 nm).

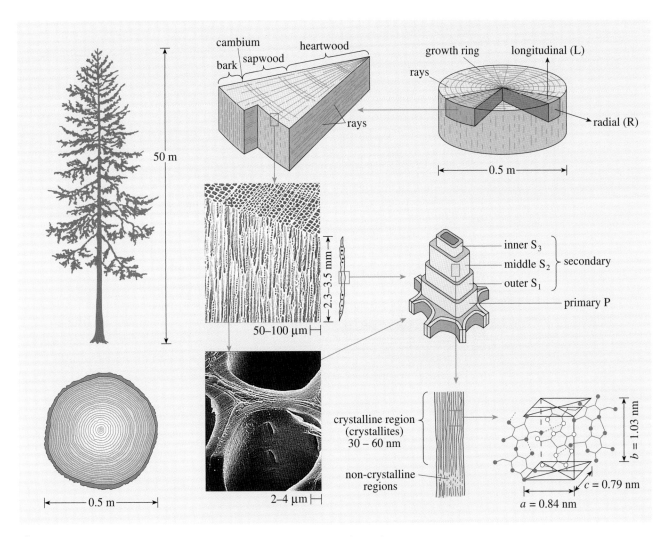

Figure 4.16 Scale and nature of structural features in a softwood

The bulk of a tree is made up of dead cells, the live, growing region being confined to the *cambium* immediately beneath the bark. Wood, of course, displays the well-known, annual pattern of growth rings. The cell wall structure contains about 40–50% by weight of cellulose and 20–25% of amorphous materials known as hemicelluloses. The remaining material is largely composed of resinous *lignin*.

A horizontal slice through the trunk reveals the annual growth rings, and removing a 'pie-slice' from this and examining it at higher magnification shows that the main structural elements are an array of hollow, vertical cells, known as *tracheids*, and these are connected by a much smaller number of horizontal, radial cells located in the *rays*. Tracheids have aspect ratios of the order of 100, and a closer look at their cell walls shows that these have a layered structure as the alignment of the cellulose molecules varies through the cell wall.

As mentioned earlier, since wood is a natural material the properties obtained from a set of samples of any one species are likely to show a much greater variation than several samples taken from a synthetic material. Standard deviations of 20% are typical. Density is one important variable, and both this and other properties are significantly affected by the moisture content. A further complication is that wood, like most polymers, is a viscoelastic material – it experiences creep under load and recovers rather more slowly after removing the load.

4.2 Modelling the properties of wood

With the above factors in mind, let's look at the properties of some hardwoods and softwoods. One feature of woods is that, despite the large variation between and within species, the composition, structure and properties of the cell *walls* remain very much the same. Typically they have a density of 1500 kg m^{-3} and Young's modulus of 35 GN m^{-2} and 10 GN m^{-2} parallel to and transverse to the cell axis, respectively. This suggests that the overall density of the material relative to that of the cell wall must be an important determinant of the mechanical properties.

Figure 4.17 shows the Young's modulus values of some woods plotted against their density. The two series of data represent the modulus parallel to (E_L) and perpendicular to (E_R), the direction of growth of the original timber. The numerical data on which the graph is based are given in Table 4.5. You can see from the lines drawn on the graph that the parallel modulus is roughly proportional to the density of each wood, whereas the perpendicular modulus is more nearly proportional to the square of the density. But even the least anisotropic of the woods still shows a five-fold difference between its modulus in the two directions. Hence at least some of the attractions of the wood composites you saw earlier.

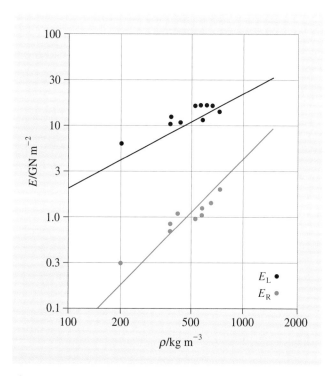

Figure 4.17 Young's modulus *v.* density for a selection of woods. Both axes have logarithmic scales

Table 4.5 Typical density and Young's modulus values of some woods

Species	ρ/kg m^{-3}	E_L/GN m^{-2}	E_R/GN m^{-2}
Hardwoods:			
balsa*	200	6.3	0.30
mahogany	440	10.2	1.13
walnut	590	11.2	1.19
birch	620	16.3	1.11
ash	670	15.8	1.51
oak	690	13.6	1.28
beech	750	13.7	2.24
Softwoods:			
Norway spruce	390	10.7	0.71
Sitka spruce	390	11.6	0.90
Scots pine	550	16.3	1.10
Douglas fir	590	16.4	1.30

* Note that, contrary to what you might expect, balsa is correctly classified as a hardwood.

Exercise 4.10

Use the merit indices we developed earlier to find the top three types of wood for making a lightweight, solid rectangular-section floor joist for each of these two conditions:

(a) where the width of the joist is fixed but the depth may be varied;

(b) where the depth is fixed but the width may be varied.

How realistic are these merit indices for choosing structural materials for the average domestic house? What other index might you use?

SAQ 4.9 (Outcome 4.5)

The floor joists in my old house in the south of England are made of oak, in common with many of the older buildings in the same area. In a modern house anywhere in Britain they would generally be made from something similar to Sitka spruce. What factors do you think are at play that might explain this difference?

How does this relate to the merit index exercise you have just done?

5 CONCRETE

5.1 The nature of concrete

Concrete is a composite material. It is a mixture of different materials – cement and aggregate – each of which adds its own particular characteristics to the mix. What is more, the processing of concrete involves both physical and chemical changes in some of the ingredients, making it a fascinating and complicated material to study. ▼The outline chemistry of lime-based cements▲ provides a very brief overview of this subject.

▼The outline chemistry of lime-based cements▲

All *cement* starts with some form of naturally occurring mineral with a high content of limestone (a form of calcium carbonate). On heating in a furnace, calcium carbonate turns into quicklime (calcium oxide) and releases carbon dioxide. Addition of water to quicklime 'slakes' it to produce slaked lime (calcium hydroxide). Slaked lime rather slowly absorbs carbon dioxide from the atmosphere. This in turn forms calcium carbonate, the mineral with which the process began. This is the upper cycle in Figure 4.18.

The use of building materials based on pure lime has been widespread throughout history in areas where clean limestone or chalk (simply another form of limestone) are easily extracted. But the slow rate of 'carbonation' of slaked lime and the fact that it has to dry out before the reaction can start are distinct limitations on the kinds of structure that can be built this way, the rate at which they can be built, and their location.

Changes to the chemistry are, however, possible. The Romans, for example, used to add to their lime volcanic rock containing significant quantities of aluminium and silicon in a highly reactive form. In Britain and France, some naturally occurring limestone deposits contain iron, aluminium and magnesium compounds as well as a proportion of sand (silicon dioxide). That now ubiquitous material, Portland cement, is made by firing nearly pure limestone with clay, which contains aluminium, silicon and iron.

The inclusion of other minerals in this way, either before or after firing, provides alternative chemical reactions to the simple carbonation of slaked lime. In particular, the aluminium, silicon and iron compounds combined with the calcium react at various rates with water – a range of so-called 'hydraulic' reactions – to create a mixture of minerals that mimic many of the properties of naturally occurring building stones. In fact, the name Portland cement comes about because of its similarity in appearance to the limestone quarried on the Portland peninsula on the south coast of England.

This chemistry is summarized in Figure 4.18. It is still the subject of much research because of its complexity.

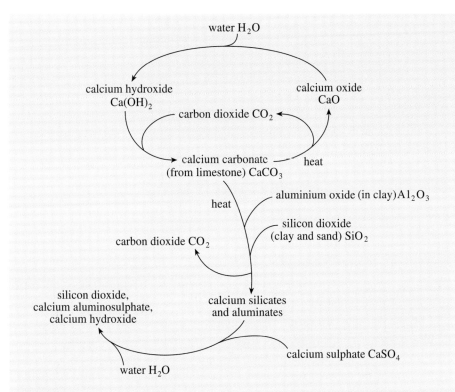

Figure 4.18 A schematic representation of some of the chemistry of cement materials. (Note that many aspects of the chemistry of cement are not shown.)

The advantages of hydraulic materials over non-hydraulic lime can be presented as:

- hydraulic materials harden on mixing with water, rather than on drying out;

- rates of reaction are quite rapid, allowing buildings to be erected much faster;

- hydraulic materials can be used in bulk (carbon dioxide penetrates only slowly into thick sections).

Portland cement has some distinct disadvantages compared to lime, particularly in the repair and restoration of buildings originally constructed with lime. But that's another story.

Lime-based materials that show little or no hydraulic reactivity develop their strength only very slowly with time. Even Portland cement takes hours to begin to harden, and is not normally regarded as fully serviceable within about a month of processing. Materials are therefore added to the cement to provide some strength during the early stages of application. They also continue to contribute to the strength of the resulting composite material throughout its life. Such materials, collectively referred to as *aggregates*, range from sand and crushed brick through to gravels, crushed rock and naturally occurring stones and pebbles. Cement–sand mixtures are traditionally called *mortars*, but mixtures with a significant proportion of large aggregate are known as *concrete*. Figure 4.19 shows an optical micrograph of a thin slice through a sample of concrete.

Figure 4.19 Optical micrograph of concrete

In the micrograph the discrete shapes are all particles of aggregate, and you can see that they come in a wide variety of shapes and sizes. The millimetre-sized bright particles are sand. The larger grey particles are various types of gravel and stone, and you can see that some of them have some internal structure, such as the large particle to the far left. The aggregate particles are all surrounded by darker grey material that is the cement 'matrix'. You can begin to see the difficulty of amalgamating such a complicated mixture when you look at the bright area that divides the image roughly down the centre. This is where the cement and aggregate have not mixed properly, and it would represent a line of weakness in the concrete. The black shapes are pores in the concrete left behind from air incorporated in the mix or from the evaporation of excess water after the concrete set.

The cement has two distinct functions to perform in concrete:

- when mixed with water, it forms a semi-liquid 'slurry' that allows the cement/aggregate/water mixture to be processed by casting into a confined space;

- once hardened it serves to hold the strong, hard aggregate particles together, allowing loads to be transmitted through the composite material.

Hydrated cement alone is not a useful building material, neither is raw sand nor gravel; but in combination they are useful. But there is a further important aspect to including a high proportion of aggregate in concrete. Cement is expensive compared to the other ingredients, largely because of the energy required in its manufacture – typically about a tonne of coal or its equivalent is consumed for every five tonnes of cement. Despite this, when ▼**Weighing the costs**▲ cement and concrete compare favourably with other materials in both money and energy costs. It still remains, however, an objective when specifying concrete mixes to produce a desired strength at lowest cost, and this means minimizing the content of the most expensive ingredient, the cement. The corollary to that goal is that the amount of aggregate must be maximized, so we shall look more closely at that ingredient next.

5.2 Controlling the properties of concrete

Concrete depends for its strength and stiffness on the properties both of the cement matrix and the aggregate, as you will see shortly. Working on the principle that the prime purpose of the cement is to hold the aggregate particles together, the key to creating a concrete with a desired set of properties lies in the choice of aggregate – not just its composition, but also its size and shape.

Aggregate for concrete usually consists of a mixture of sand, which has a mean particle size of less than 2 mm, and crushed rock or gravel, in which the mean particle size is greater than 2 mm. If the concrete is to act as more than a loose pile of sand and gravel, the mix must contain sufficient cement paste to coat all the aggregate particles and fill the spaces between them, as you saw when considering Figure 4.19.

The amount of cement paste needed can be reduced by using what's known as a graded aggregate. The usual ratio of aggregate to cement is about five volumes of graded aggregate to one volume of cement. The aggregate contains particles of a range of sizes. The small particles fit in the spaces between the large particles, and the cement is required only to flow into the remaining spaces (Figure 4.20). Typically the proportion of sand to gravel is about 60:40.

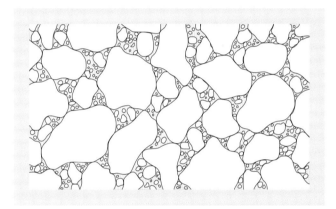

Figure 4.20 Schematic of graded aggregate

▼Weighing the costs▲

It is instructive to compare cement and concrete with some of the other materials used in construction. Here, as well as money costs, we are going to look at energy costs. Table 4.6 sets out the relevant data, as well as giving the energy q required to produce unit volume of the material.

What shows up immediately is how low both the cost per unit mass and the energy per unit volume are for cement and concrete compared with the other synthetic materials in the table. Softwood, the one natural material in the table, has an even lower energy content because it does not consume much energy in processing.

Of more interest perhaps, is to establish what it costs in money or energy terms to buy unit stiffness or unit strength. Because of the combination of low money cost and low energy requirement, concrete shows up very well in such comparisons, despite its low value of tensile strength and its relatively low value of Young's modulus. These, coupled with its mouldability, explain why it has become such an important construction material.

SAQ 4.10 (Outcome 4.5)

Suppose an application requires a short column (i.e. buckling is not an issue) of length L and cross-sectional area A to carry a given compressive load, F. Write down expressions, in terms of these parameters and those in Table 4.6, for each of the following:

(a) the stress in the column at failure;

(b) the cost of the column;

(c) the energy content of the column;

(d) the compressive stiffness of the column (force per unit deflection).

Now use these expressions with the merit index approach as developed in Section 3.5 to decide which material in Table 4.6 offers:

(e) the column with the lowest cost that will not fail under load;

(f) the column that requires the least energy to produce the material;

(g) the column with the lowest cost per unit of stiffness under load.

Table 4.6 Typical cost, property and energy data for some materials used in construction, for SAQ 4.10

Material	Relative cost per unit mass c/kg^{-1}	Density $\rho/\text{kg m}^{-3}$	Elastic modulus $E/\text{GN m}^{-2}$	Strength $\sigma_p/\text{MN m}^{-2}$	Energy cost $q/\text{GJ m}^{-3}$	Merit index for part (e)	Merit index for part (f)	Merit index for part (g)
Cement	100	2400	30	16 *	12			
Concrete	50	2400	30	26 *	8			
Aluminium	2600	2800	71	200	500			
Mild steel	600	7860	210	350	300			
Softwood	500	400	11	50	3			
Window glass	1000	2500	71	50	50			
UPVC	1200	1450	3.0	50	70			

* Values in compression

Most aggregates are obtained from naturally occurring sand and stone deposits. The type of stone from which they are derived varies, as does the size and shape of the particles. Shape is particularly important in determining the strength and the ease with which the concrete mix can be worked. Angular stone particles, such as might be obtained from crushing quarried stone, produce a concrete of high strength, but one that is difficult to work with; the interlocking of the angular particles restricts the flow of the mix. River gravels comprise smoothly rounded stones which flow easily over each other, but produce a lower-strength concrete.

The requirement that the cement paste should both coat the aggregate and fill the spaces between the aggregate particles means that for a given aggregate there is a certain ratio of cement to aggregate that provides the optimum mix. Less cement means fewer voids will occur in the mix; more cement than the optimum means the mix is more expensive with no advantage in terms of properties. This also affects the water/cement ratio and, hence, how easy it is to work with the concrete after mixing (i.e. the 'workability' of the mix).

The practicalities of casting the liquid concrete figure largely in designing and selecting a mix. A pure cement paste made with a water/cement ratio of 0.4 will not flow easily. Hence a concrete mix containing cement with this ratio will be unworkable – it just can't be poured. In the same way as cement, concrete with a high water/cement ratio will have a lower strength than concrete made from similar aggregate, but with a lower ratio. The extra water required to make the mix workable over and above that required to hydrate the cement increases the porosity of the concrete. It is possible that quite large holes can occur if the concrete is not properly compacted. One way round this is to use flow-enhancing admixtures (plasticizers) and to keep the water/cement ratio low. Similar problems occur in selecting aggregate sizes. If thin sections are required, then large aggregate particles will hinder easy flow and can cause large voids.

Assuming that the cement and aggregate are well specified and mixed, the models that I developed in Section 3.3 can be used to quantify the contribution of each component of the mix to the overall performance of the composite material.

SAQ 4.11 (Outcome 4.4)

Which of the models derived in 'Combinations of materials' (Section 3.3) do you think is most likely to fit the behaviour of concrete, and why?

Figure 4.21 shows some data for real concrete, confirming that it fits quite closely to the homogeneous stress behaviour described, at least over a practical range of compositions. Table 4.7 gives the density and Young's modulus for the components of concrete.

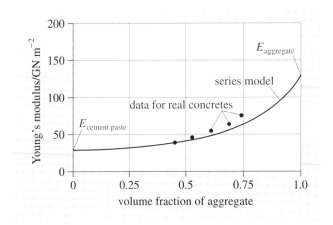

Figure 4.21 Application of homogeneous stress composite model to the elastic modulus of concrete

Table 4.7 Typical properties of cement and aggregate

Material	Density kg m^{-3}	Young's modulus GN m^{-2}
Cement	3000	30
Sand	2650	90
Granite	2650	70
Limestone	2400	20
Sandstone	2300	10

While analyses of this type are useful for estimating the modulus of a composite, estimates of its strength depend on the point at which the material ceases to behave elastically. This might occur because of the failure of one component by yielding or fracture, or by the breakdown of the interface between components. Concrete is specified in terms of its compressive or crushing strength, expressed as its ▼**Cube strength**▲.

The correlation between particle shape and strength that I mentioned earlier perhaps gives some indication of the nature of the bonding between the cement paste and the aggregate. Mechanical interlocking of the cement paste into the surface irregularities in the aggregate particles must play a significant role in determining the strength of concrete.

It is generally found that the mechanical failure of concrete involves cracking of the aggregate–cement interface. At first, that might seem like a serious disadvantage, since the effectiveness of the aggregate as reinforcement for the cement depends on a good connection between the two. But, in fact, failure of the interface is one of the most important mechanisms by which two brittle materials can nonetheless be combined to give a composite that absorbs a substantial amount of energy during fracture – in other words, a tough material.

However, because concrete is made up of two brittle components of similar modulus, it can only be used safely in conditions where it is subjected to *compressive* stress. Both constituents fail at similar strains, so once the critical strain is reached in tension, a crack will propagate rapidly. For this reason alone, concrete is never used on its own where there is the slightest risk of applied loads giving rise to *tensile* forces within the material. Instead, it is almost universally combined with steel to create 'reinforced' concrete; so steels are what we shall look at next.

▼Cube strength▲

In the United Kingdom, the standard quality control test for the strength of concrete involves measuring the load at which a 100 mm cube of the material is crushed. Figure 4.22 shows a (now obsolete) standard test which used a six-inch cube. The piece of concrete at the bottom left is an enlargement of the six-inch cube after failure in compression, and displays the classic characteristics of compression failure where wedges of material break away from the sides of the test piece at an angle of 45° to the direction of the applied load. If a series of specimens is cast at the time the concrete is poured, they can be used to monitor the hardening with time of concrete in a structure.

Figure 4.22 The remarkable strength of concrete

The stress produced by dividing the crushing load by the cube's cross-sectional area is known as the *cube strength*. It is *not*, however, an intrinsic material property (such as, say, yield strength or fracture toughness), and the *in situ* strength of cast concrete is typically only about 65% of its cube strength.

Concrete *grades* are specified in terms of their cube strength 28 days after mixing, measured in MN m^{-2}. Grade 20, corresponding to a cube strength of 20 MN m^{-2}, is just a space-filling grade and grades 30 and 40 tend to be the popular ones. Grades with strengths higher than 40 MN m^{-2} are expensive and are normally used only in applications which justify the extra cost.

Exercise 4.11
What are the cube strength and grade of a concrete whose compressive failure load was 30.6 tonnes in a standard UK test?

6 STEELS FOR STRUCTURES

There is an almost bewildering number of steels available to the engineer, with a huge variety of mechanical property profiles. The only certainty is that, as usual, increased performance is closely associated with increased cost.

The vast array of different steels calls for a classification system to enable engineers and designers to specify the type of steel they require for any application. A number of such systems exist, but the one used in the UK gives steels designations such as **S 410 N JR H**. Each part of the designation has a different meaning, relating to its intended application, mechanical properties, metallurgical condition and so on. To make sense of the reasons behind this unusually extensive range of choice involves significant understanding of the underlying metallurgy of steels. This is outside the scope of this course. Nevertheless, I shall endeavour to introduce you to the main classes of structural steels and the basics of the fundamental differences between them. Such knowledge is essential to the design and construction of durable and efficient engineering structures such as those described earlier in this block.

The word 'steel' conjures up all sorts of associations in common use. We describe someone as having 'nerves of steel' or by association as being 'hard as nails'. The image of steel manufacture as being critical to manufacturing industry has relaxed somewhat in recent years, but steel is still used in quantities that dwarf those of all other metals. There are two main reasons for this popularity. Firstly, as noted earlier, steels can be made to have mechanical properties suitable for a wide range of applications, as may be seen from Figure 4.23.

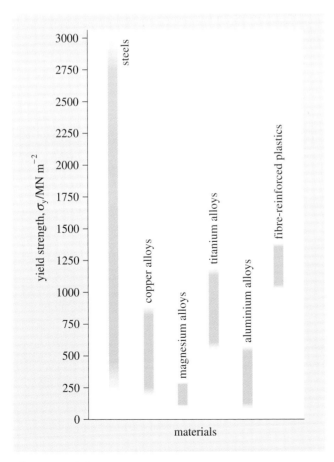

Figure 4.23 Comparative strength ranges of some materials

Secondly, iron is readily available and relatively inexpensive. Exploitable iron ores such as magnetite, Fe_3O_4, and haematite, Fe_2O_3, are thought to be sufficiently abundant to produce over 7×10^{10} tonnes of steel per year! This, coupled with the fact that the oxides of iron can be easily reduced by carbon at temperatures just above 1000 K, historically led to the construction of large steel-making plants in most industrial countries and a surviving competitive international market in most steel grades.

To appreciate how such a variety of properties can be obtained from what appears to be the same material, we need to start with the atomic-level structure of plain carbon steels. Then I shall show you how this interacts with iron's allotropic behaviour to provide almost unparalleled control over microstructure.

Allotropy is discussed in the Appendix to this block.

Before reading Section 6.1, you should familiarize yourself with the Appendix, which deals with phase diagrams, solid solubility, eutectic and FCC/BCC crystal structures. Some of this may be revision of topics you have studied in previous courses, such as T173, or elsewhere.

6.1 Steels as alloys

There are two ways in which alloying elements dissolve into a parent crystal matrix. In the case of copper–nickel alloys, the copper and nickel atoms are roughly the same size. Simply substituting one for the other in the lattice happens quite readily, allowing these alloys to display total solid solubility.

Where the atoms are rather different in size, the lattice has to 'give' a bit to allow the substitution of one type of atom by another. This is the situation with tin–lead alloys and the result is a separation of phases – lead dissolved in tin, and tin dissolved in lead. But, even so, this is still a case of substitution of one atom for another. Both copper–nickel and tin–lead alloys are examples of *substitutional solid solutions*.

There is another possibility, though, when the atoms of the alloying element are significantly smaller than those of the lattice: the dissolved atoms can fit into the spaces between the atoms of the crystal, known as interstices (pronounced 'in-ter-stis-sees') to create an interstitial (in-ter-sti-shal) solid solution as shown in Figure 4.24.

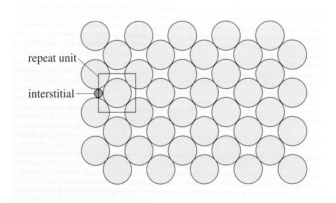

Figure 4.24 An interstitial solid solution accommodates the dissolved atoms in the gaps between the atoms of the parent crystal

This is what happens when carbon is added to iron. The carbon atoms are so small that they can fit into the interstices in the FCC or BCC iron structures. The carbon atoms are not so small, however, that they can slip into the structure without having some effect on the behaviour of the metal. Let's examine why.

Looking at these diagrams it is easy to make the mistake of thinking that atoms behave like hard spheres. It's not quite that simple. Nevertheless it is possible to estimate how much space exists between atoms in a crystal (in other words the dimensions of the interstices) and to suggest how this affects the ability of the crystal to absorb 'foreign' atoms. The largest space in an FCC iron lattice works out at about 0.1 nm across. That in a BCC iron lattice is only 0.07 nm. But carbon atoms are more like 0.14 nm in diameter. So dissolving carbon into the interstices in the iron crystal lattice results in a quite severe distortion of the regular array of iron atoms (Figure 4.25). The result is that the maximum amount of carbon that can dissolve in BCC iron at room temperature is around 0.005% by weight. In FCC iron, with its larger interstices, that increases to more than 1%. Remember that iron at room temperature is in the BCC form, and it changes to FCC only above 910 °C.

Because an atom of iron has a mass of over 4.5 times one of carbon, these solubilities work out at approximately one atom of carbon for every 20 atoms of iron in the FCC form, compared to one for every 4000 in the BCC form.

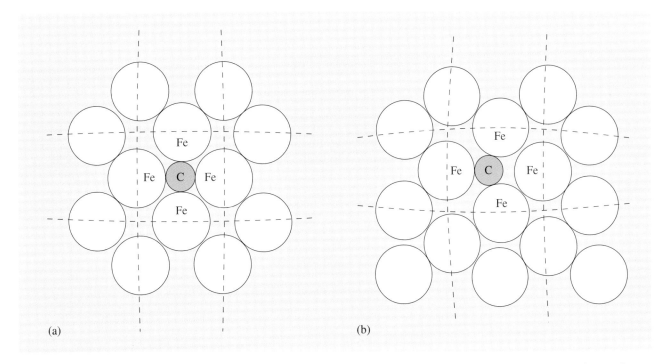

Figure 4.25 The interstices in both FCC iron (a) and BCC iron (b) are only large enough to accommodate carbon atoms if there is some distortion of the lattice

To begin to appreciate the consequences of this difference in the solubility of carbon in iron it helps to look at a portion of the equilibrium phase diagram that describes the iron–carbon system. The majority of engineering steels contain less than 1% C, so the microstructures and properties of all steels are dominated by what may or may not be shown by this part of the iron–carbon phase diagram (Figure 4.26 overleaf).

If you look at the white area of the diagram you can see that it has the same basic shape as the Pb–Sn eutectic phase diagram (Figure A.9 in the Appendix). The Pb–Sn phase diagram exhibits a eutectic point in the liquid–solid reaction, whereas in Figure 4.26 the phases both above and below the *eutectoid point* are all solid. The eutectoid point is distinguished by its *eutectoid temperature* and *eutectoid composition*.

A eutectoid point is the same as a eutectic point but in the solid state.

You can see from the phase diagram that the phases present between room temperature and 1200 K are α, γ and Fe_3C. (There is also a β phase but it looks and behaves so much like α that it is not normally included in the diagram.) Each of these three phases has its own name:

- α is called 'ferrite'
- γ is called 'austenite'
- Fe_3C is referred to as 'cementite'.

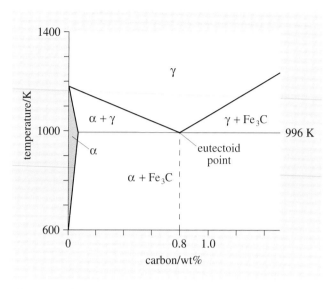

Figure 4.26 Part of the iron–carbon phase diagram

Exercise 4.12

(a) Write next to the α and γ labels on the phase diagram the crystal structures of the respective phases and their metallurgical names.

(b) For the eutectoid composition (0.8% carbon), what phases are present just above the eutectoid temperature (996 K)?

(c) For the eutectoid composition, what phases are present just below the eutectoid temperature?

(d) What is the maximum solubility of carbon in iron at 1100 K?

(e) What is the maximum solubility of carbon in iron just below the eutectoid temperature?

(f) What do you suppose happens to all the carbon that was dissolved in the γ phase on cooling slowly through the eutectoid temperature?

You have now encountered one of the enormous obstacles to learning about steels – the vocabulary that has developed around them. You may find the ▼Glossary of steel terms▲ useful as a reference as you study the next few sections.

▼Glossary of steel terms▲

Austenite (γ): The FCC form of iron and of solid solutions based on it. In pure iron, it is stable between 1180 K and 1660 K.

Cementite (Fe_3C): A chemical compound with the precise composition given by its chemical formula. By weight it contains 6.67% carbon. It has a complex hexagonal crystal structure. It is very hard and brittle.

Eutectic point: The point in a phase diagram where a single liquid phase transforms to a two-phase solid solution on cooling past a specific temperature.

Eutectoid point: The point in a phase diagram where a single solid phase transforms to a two-phase solid solution on cooling past a specific temperature.

Ferrite (α): The BCC form of iron and of solid solutions based on it. In pure iron, α ferrite is stable up to 1180 K.

Graphite: The most stable form of carbon in the Fe–C system, but usually found only in the cast irons.

Martensite: Non-equilibrium microstructure formed by cooling austenite rapidly. The entrapped carbon distorts the crystal structure to give a tetragonal lattice. It can be very hard and brittle.

Pearlite: A microstructure formed at the eutectoid composition (0.8 wt% C), it consists of alternating layers of ferrite and cementite.

Quenching: The process of rapid cooling by plunging a heated object into a blast of air, oil, water or iced water.

Tempered martensite: A microstructure obtained by reheating martensite to a specified temperature to allow partial transformation to ferrite and cementite.

6.2 Controlling the properties of steels

The interstitial carbon atoms in steel are relatively mobile and carbon can redistribute within an iron crystal lattice quickly compared to other alloying elements. But the length of time it takes is a critical aspect of how steels behave.

Look again at the part of the iron–carbon phase diagram which is shown in Figure 4.26. If a steel component is cooled slowly from the austenite phase field, where all the carbon is held in solid solution, then the microstructure of the steel changes from consisting solely of austenite grains to a mixture of ferrite and cementite. This means that the majority of the carbon atoms must leave their random locations in the interstices in the FCC austenite and form ordered combinations with iron atoms as cementite – there simply is not enough space for most of them in the interstices of the BCC lattice.

Although carbon atoms are quite mobile in steel, this redistribution by diffusion still takes some time and their speed of travel is very dependent on temperature. The atoms possess much more kinetic energy at higher temperatures, so the speed at which they diffuse through the crystal is higher. If you cool a steel from the austenite phase field fast enough (in a process known as 'quenching'), some of the carbon atoms do not have time to diffuse out to form cementite but are trapped in the BCC ferrite lattice. This distorts it badly, to give the atomic arrangement shown on the left in Figure 4.27.

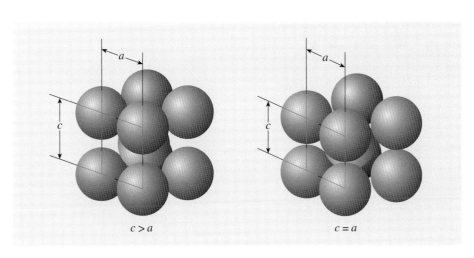

Figure 4.27 Trapping a few per cent of carbon atoms in the iron on rapid cooling forces the previously BCC crystal (right) into an unstable body-centred tetragonal (BCT) structure (left). The distances c and a are measured between the geometric 'centres' of the atoms

Material with this distorted structure is called *martensite* after the metallurgist who identified it. The distortion makes the steel very hard and brittle – an effect that has been known for hundreds of years. However, it is only in the last century or so that it has become sufficiently well understood to allow metallurgists to design and manipulate steel microstructures and properties at will. Unlike the forms of iron and iron–carbon alloys that I have introduced so far, martensite is a *non-equilibrium* structure. It can only form under particular conditions – specifically rapid cooling, or *quenching*. I want to look next at how to manage this quenching process in order to produce steels with a given set of properties.

Quite often we wish to take advantage of the characteristics of steels to be hardened through the formation of martensite. The brittleness is frequently a disadvantage, and I shall come back to that shortly. To control the process we need to know quite a lot about the times and temperatures of any quenching procedure in relation to the behaviour of a steel of a given composition. For this we can use *time–temperature transformation diagrams*, usually known as 'TTT curves'. The TTT curve for a plain carbon steel is shown in Figure 4.28.

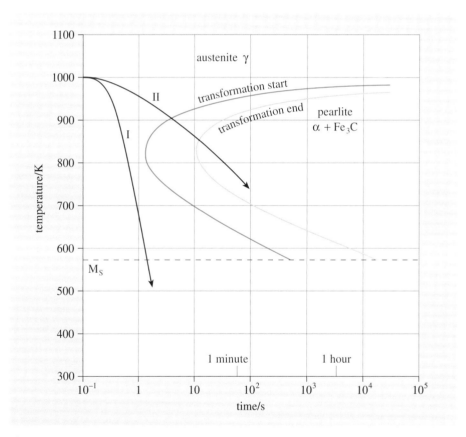

Figure 4.28 Time–temperature transformation diagram for a plain carbon steel (time plotted on a logarithmic scale)

You can see that the vertical axis is temperature and the horizontal axis is time on a logarithmic scale. To the left-hand side of the diagram and across the top you can see a region labelled 'austenite γ'. Towards the right-hand side you'll see the 'pearlite $\alpha + Fe_3C$' region. The two regions are separated by a pair of curved blue lines. The left-most curved line represents the time, at a given temperature, that transformation from austenite to pearlite starts. The right-most one marks out the time at which the transformation is complete. The area at the bottom of the graph below the dashed line represents the time–temperature regime in which martensite can form.

To make any sense of such a TTT curve, we need to plot on the diagram the 'thermal history' or 'heat treatment' of a component made from a steel with this composition. Take, for example, the black line labelled 'I' on Figure 4.28. This represents a sample that has been held for a long time at a temperature around 1000 K. (Because it has been allowed to equilibrate in the austenite region, it is said to have been 'austenitized'.) Its heat treatment then consists of a very rapid quenching to something close to room temperature. You can see that the time–temperature curve for this sample passes entirely through the γ region until it reaches the martensite region just below 600 K, labelled M_s in Figure 4.28. The result will be pure martensite, giving a very hard, brittle component.

Now look what happens if the component is cooled rather more slowly. The heat treatment shown by line II demonstrates a lower rate of heat loss than I. Instead of transforming straight from austenite to martensite, the sample now crosses the boundary into the pearlite zone. This transformation starts at about 3 seconds after the beginning of the quenching cycle and is complete in about 11 seconds. Since pearlite is a stable microstructure at room temperature, no further transformations will occur and the sample will be fully pearlitic, giving a relatively soft, highly ductile steel.

In practice it is very difficult to quench such a steel to obtain pure martensite, especially in a thick section. Look at the times involved. To avoid the 'nose' of the TTT curve completely the temperature of the centre of the thickest part of the component being quenched must cool from over 1000 K to below 800 K in under a second. So it's quite conceivable to produce a thick section with a soft, ductile core of pearlite surrounded by a hard, brittle surface of martensite. That may be a highly desirable outcome in some circumstances but it is not without its complications. The most significant of these is that the different rates of contraction of the various parts of the component on quenching can cause the martensitic surface to crack.

Fortunately for the steel metallurgist, the formation of pearlite, as defined by the position of the nose of the TTT curve, can effectively be delayed by adding alloying elements to the steel. This is because the alloying elements want to partition themselves between the two phases in the pearlite, some preferring the ferrite and some the cementite. This alloy redistribution slows the rate of transformation from austenite to pearlite. Thus, an alloy steel containing 2% manganese and 0.002% boron has a TTT curve as shown in Figure 4.29 (overleaf).

The nose of the curve now occurs at about 20 seconds, which means that quenching can be done more gently and much bigger sections can be heat-treated. The degree to which the pearlite reaction is delayed is termed *hardenability*. In practice, bars of standard size are quenched under known rates to produce data describing the hardenability of a given steel.

A steel component with a purely martensitic microstructure is of very limited use because of its extreme brittleness; standard techniques are used for ▼Assessing steel toughness▲. But the phase diagram (Figure 4.26) gives us a clue as to how we might apply further heat treatment procedures to impart a degree of ductility to the alloy. When the steel is heated back up, the increased thermal energy allows the martensite to begin to transform to the equilibrium condition of $\alpha + Fe_3C$. The rate of transformation depends on temperature, so the metallurgist can control the amount of martensite

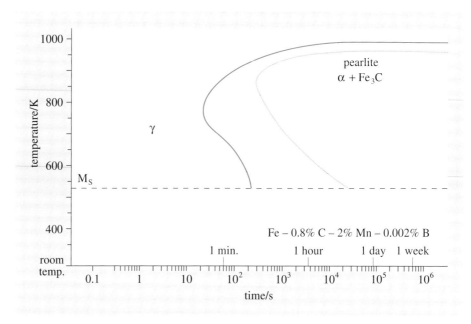

Figure 4.29 TTT diagram for a low-alloy steel

transformed using a combination of time and temperature. This process of reheating to a specified temperature for a given time is called *tempering* and the new microstructure – a halfway house between martensite and pearlite – is referred to as 'tempered martensite'. The aim of tempering martensite is to retain as much as possible of the hardness but to introduce some ductility.

▼Assessing steel toughness ▲

The majority of steels become brittle if you cool them down to a sufficiently low temperature. More worryingly, some steels can become brittle at or near room temperature. This can have severe consequences if the steel is used for a structural purpose.

An example of this is provided by the Liberty ships, Figure 4.30. The Liberty ships, produced in great numbers during the Second World War, were the first all-welded ships. A significant number of them failed by catastrophic fracture. Fatigue cracks nucleated at the corners of square hatches and propagated rapidly by brittle fracture. In earlier ships, the riveted plates acted as natural crack arresters, but these were absent in the all-welded Liberty ships. The problem was solved by improvements in ship design and steel quality.

Furthermore, under conditions of very sudden 'shock' loading, some steels behave very differently from what you might expect from the results of a simple tensile test. So some steels possess a high tensile strength but are weak under impact. We assess this tendency using an impact test, and the method used in the European standards (and defined in the UK by BS EN 10045) is the Charpy impact test. In this test a notched bar specimen of the material is broken by the swinging of a weighted pendulum, and the energy absorbed by the specimen is measured. A brittle failure will absorb much less energy than one in a tough material.

Figure 4.30 A Liberty ship split in two

A typical Charpy test arrangement is shown in Figure 4.31. The pendulum 'bob' in this case is shaped to provide a hammer-head impact on the opposite face to the notch.

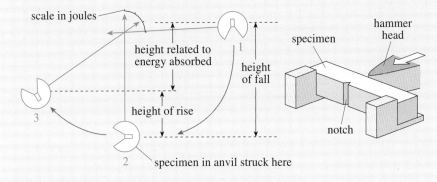

Figure 4.31 Schematic of the Charpy impact test

A complex three-dimensional stress system develops when the notched specimen is loaded, which results in an increase in the material's yield stress and a drop in its ductility. This effect, when combined with the very high deformation rate imposed by the impact, makes the test results indicative of how the material will behave under the most severe conditions likely to be encountered in service.

The relative severity of the test is demonstrated by the fact that a typical mild steel will not exhibit brittle failure in a tensile test until about –150 °C, but can show brittle behaviour at temperatures near 0 °C in a Charpy test. Clearly, if a steel is going to become brittle at a given temperature it is important to know about it, and a series of impact tests at different temperatures will reveal the temperature at which the behaviour of a given steel changes from ductile to brittle. An example test result is shown in Figure 4.32.

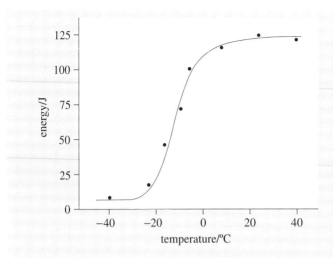

Figure 4.32 Ductile–brittle transition temperature for several samples of the same steel; brittle behaviour absorbs much less energy than tough behaviour

The so-called ductile–brittle transition temperature (DBT) is found to be highly dependent on the quality and 'cleanliness' (freedom from impurities) of the steel. Thus the way standards deal with this issue of different quality steels is by specifying a Charpy toughness – the minimum energy that a standard Charpy test specimen must absorb during a Charpy impact test. Those thinking of travelling on car ferries during the winter will be relieved to know that modern shipbuilding steels are of high quality and have ductile–brittle transition temperatures many tens of degrees below those of the Second World War steels used to construct the Liberty ships.

You will notice that the definition of Charpy toughness as the energy that a standard test specimen must absorb during an impact test is not totally consistent with the more precise definition of toughness given in Section 3.4. The Charpy impact test is only one way of assessing the toughness of steel. Indeed, there are other, generally more expensive, tests that provide a much better assessment of the toughness of a given material.

The hardness (and strength) of martensite increases markedly with the carbon content of the steel in which it is formed, as can be seen from Figure 4.33. But the low toughness of this very hard martensite, particularly at temperatures below room temperature, usually makes it unsuitable for structural use. Structural steels are therefore most often low-alloy steels (which equates to low cost) in a 'quenched and tempered' metallurgical condition.

Quenching and tempering to modify crystal structure are not the only procedures used to alter the behaviour of structural steels. Another very common goal is to reduce the average grain size within the metal – grain 'refining'. This can provide a slight increase in both strength and toughness. Grain refining involves first austenitizing the steel at as low a temperature as possible, when new, smaller grains of austenite form throughout the metal, and then cooling slowly into the $\alpha + Fe_3C$ region. Grains of ferrite and pearlite form at many different sites within the steel to produce a grain structure that is considerably finer than that of the starting material.

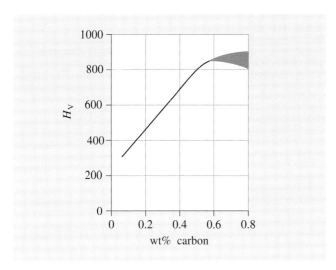

Figure 4.33 The variation of the hardness of martensite with the carbon content of a steel

An alternative route to a similar structure is the process known as 'normalizing'. After austenitizing, the steel is removed from the furnace and allowed to cool in air. The changes are the same as in grain refining, but the shorter time spent at high temperature (where grains grow over time rather like the soap bubbles in a bath) and the faster cooling rate combine to give a slightly finer structure. Normalizing is also a cheaper process as less time is spent in the furnace. But it can only be used for fairly uniform sections where the distortions that result from differential cooling are minimized. The term 'normalized' thus describes both a process and a metallographic microstructure when applied to a structural steel. Figure 4.34 shows the characteristic fine-grained microstructure of a normalized structural steel.

Figure 4.34 Microstructure of a normalized structural steel

Finally, if a structural steel contains other specialized elements such as niobium or vanadium, then heat treatment can be used to precipitate a fine dispersion of hard carbide particles throughout any ferrite in its microstructure. Such steels are termed 'precipitation hardened' as they offer increased strengths compared to low-carbon steels.

6.3 Choosing the right steel

The actual steel an engineer will select for a given structural application will depend mostly on two factors:

- the mechanical properties required (remembering that density and Young's modulus are essentially the same for all steels, that still leaves choices for strength and toughness);

- the suitability of the steel for the intended method of manufacture and assembly of the structure.

Let's look at the second of these factors in a little more detail.

If the structure is to be bolted or riveted together, then assembly methods are generally not going to influence the choice of steels. Most steels are likely to be fit for the purpose. If, however, a welded construction is planned, then the way that different steels respond to welding becomes important. The ▼**Weldability of steels**▲ is intimately connected with their hardenability and their response to the kinds of uncontrolled heat treatment that often occur during welding. The formation of martensite in the heat-affected zone (HAZ) around a weld can have dire consequences if it is not expected. Look back, for instance, to Figure 1.32 in Part 1 and think about what might happen if the cantilever beam shown there were made from steel and 'built in' by welding it to a vertical support.

SAQ 4.12 (Outcome 4.6)

Why might you worry about martensite forming in the heat-affected zone of a weld at the junction between the cantilever beam in Figure 1.32 and its support (the 'root' of the cantilever)?

If you were offered the choice of a normalized steel or a quenched and tempered steel to use for the cantilever, which would you prefer to use if it were to be welded to its support, and why?

▼Weldability of steels▲

You may well come across a description of a particular steel as 'weldable' or 'not weldable'. These terms refer to the tendency of the alloy to form martensite in the area around a weld. During welding, the surfaces to be joined are heated either very close to or above the melting temperature of the material. The heat being applied to the weld area cannot be restricted to the immediate surface layer but affects a region surrounding the weld. This region is known as the *heat-affected zone* (HAZ). The width of the HAZ and the temperature profiles across it depend on the particular welding process being used, but at least part of the material will be heated into the austenite region in the case of steels. On cooling, therefore, the austenite must transform.

Exercise 4.13

(a) List the transformations that a low-carbon steel can undergo when cooling from above 1000 K to room temperature.

(b) What process factor is most important in determining the final form of the steel for a given carbon content?

(c) What part does carbon content play in determining the microstructure of the steel?

If the cooling rate is high enough, martensite may form in the HAZ. Weldability is therefore directly related to hardenability. The weldability of any alloy composition is expressed quantitatively as its 'carbon equivalent' value – a measure of the delay in the formation of pearlite induced by the various alloying elements or, in other words, the position of the nose of the TTT curve. There is a whole range of formulae for estimating carbon equivalents, but all recognize the varying contributions of the different alloying elements to the hardenability of the alloy. One of the standard formulae used is:

$$C_{equiv} = C + \frac{Mn}{6} + \frac{Si}{24} + \frac{Cr}{5} + \frac{Ni}{40} + \frac{Mo}{4} + \frac{V}{14}$$

where each of the chemical symbols represents the content of that element in weight per cent.

Weldable steels are defined as those with less than a certain carbon equivalent, and a figure of 0.4% is widely quoted. You must appreciate, however, that this single value arises from a combination of two factors. The first is the possibility of martensite forming in the HAZ and the second is the increasing hardness and brittleness of the martensite which forms as the carbon equivalent of the steel increases. So with high carbon equivalent steels both the chances of martensite forming and its brittleness when it does form are increased, leading to a far greater risk of brittle fracture in the HAZ.

So you can see that the choice of steel for a particular structure is not simply a matter of quantifying the forces in the structure and specifying a material that can withstand the worst of them. Often the mode of construction must also be taken into account.

For the last part of this section I want to return to the point I began with – the way that structural steels are specified to make the job of picking a particular grade more straightforward.

6.4 Specifying steels

Its availability, relative cheapness and clear versatility maintain steel's popularity over other metals in most heavy industries, including civil and structural engineering. Unfortunately, this dominance of steel over other metals has led to an overwhelming number of steel grade-specifications emanating from virtually every industrialized country. Indeed, a good living can still be had just 'translating' between the various steel specifications. Over the past decade there has, however, been some effort to rationalize the situation, albeit only regionally. I will use the current European standards (enshrined in the UK in BS EN 10027) to try to produce an overview of available structural steels.

Under BS EN 10027 each steel grade is given a 'name' which is made up of a designatory letter, which is dependent on intended end use, followed by a figure for the minimum specified yield stress in MN m^{-2} and a series of optional letters which denote metallurgical condition, quality (in terms of toughness) and product form. Perhaps the best way to explain this method is to consider a fictitious steel grade, such as the one I used at the start of this section:

S 410 N JR H

The first letter, S, shows that the intended purpose is for structural use. The full range of possibilities is:

S – **S**tructural steels

P – steels for **P**ressure purposes

L – steels for **L**ine pipework

E – **E**ngineering steels

The number is the specified minimum yield stress in MN m^{-2}. So S 410 is a structural grade steel with a yield stress of at least 410 MN m^{-2}.

The next letter, N, shows that the metallurgical condition is normalized. The full list is:

N – normalized

P – quenched and tempered

A – precipitation hardened

Next, the letters JR specify the quality of the steel in terms of its toughness. In this case it would be a requirement to have a Charpy toughness of at least 27 J for a standard test specimen at room temperature. The various options provide for the same minimum Charpy toughness (usually 27 J) at different test temperatures:

JR – at room temperature

J0 – at 0° C

J2 – at –20° C

The letters K, L and occasionally G are also used in place of J to specify particular Charpy toughness requirements of a given steel grade, but there is no uniformity and the precise values are not consistent from standard to standard.

Finally, a letter or letters may be included to denote other characteristics of the steel or the form in which it is supplied. So the H in S 410 N JR H shows that the steel is supplied in a hollow section such as a tube. The letter W could be added to denote a weathering steel (a steel that has a chemical composition which limits corrosion when exposed), whilst P is used for a grade with a higher phosphorus content.

Corrosion is a topic covered in Block 5.

Overall, however, steel remains a relatively cheap, strong, stiff and highly durable material that has provided the backbone of twentieth-century industrialization and, despite recent growth in many more technically sophisticated materials, steel is likely to remain dominant in structural engineering for many years to come.

SAQ 4.13 (Outcome 4.6)

A supplier offers you I-section steel beams in a variety of sizes in steels designated:

S 355 N J0   S 275 N J2   S 410 P JR

Briefly distinguish between each of these steels according to the BS specification. Outline in a few words why you might choose each one of them in preference to the others for a given application.

---

7 REINFORCED CONCRETE

7.1 How reinforcement works

The idea behind reinforced concrete is to offset the low tensile strength of concrete by combining the tensile strength of steel with the compressive strength of the concrete to give a durable, inexpensive structural material capable of withstanding bending loads. The steel used for the main reinforcing bars is a heavily cold-worked, low-carbon ('mild') steel. The complex assemblages of reinforcing steel are a common sight on building sites (Figure 4.35). They consist of the main reinforcing bars joined by links whose function is to hold the bars in place while the concrete sets. The links are made from a softer mild steel that is easily bent to shape. The assemblies of reinforcing rods are frequently complex and the reason for this becomes a little clearer when considering the ▼Stress states in beams▲ where we can get a little closer to reality than the somewhat idealized models of beams that I have presented to you up to now.

Figure 4.35 A steel-reinforcement assembly in the process of being encased in concrete

▼Stress states in beams▲

To predict (or design) the performance of any particular structural component requires a more detailed understanding of the internal forces within the component than the simplified bending moment and shear force distributions presented in Part 1 of this block. In the case of a beam, for example, you have seen that compressive and tensile forces are induced above and below the neutral axis respectively. You have also seen that the beam experiences vertical shear forces. However, there are also horizontal shear forces arising from the bending of the beam by virtue of its finite thickness. This much is obvious if you try the 'thought experiment' suggested in Part 1 of substituting the beam with a deck of cards, but this time stacked horizontally rather than vertically. On bending (a familiar action for any card player) the cards slide over each other (Figure 4.36).

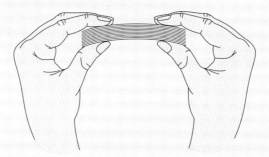

Figure 4.36 Bending a deck of cards demonstrates a shearing action

So the reality is that any point in a real structure under load will exist in a three-dimensional or *triaxial* stress state. It can be shown that it requires a total of six *stress components* to fully describe such a triaxial stress state in a theoretical cube-shaped element of a component (Figure 4.37). These are three direct (that is tensile or compressive) stresses σ_x, σ_y and σ_z and three shear stresses τ_{xy}, τ_{yz} and τ_{zx}. (The first letter of the subscript indicates the axis of the normal stress operating on the relevant plane and the second letter the direction in which the shear stress is operating.) Only three shear stresses are required because in a small cube, $\tau_{ij} = \tau_{ji}$. You'll be relieved to learn that the cases we'll investigate don't call for that level of complexity.

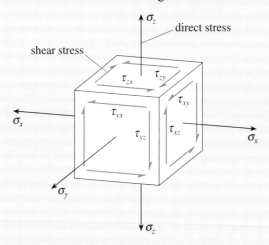

Figure 4.37 A general triaxial stress system

As the depth of a beam increases in relation to its span, shear forces become significant. They also need to be considered around the supports of longer beams. You have seen earlier how the distribution of shear forces varies depending on the manner of loading and the nature of the beam supports. All of these must therefore be taken into account in the placement of reinforcement. I do not intend to pursue the complex aspect of the placement of reinforcing rods to any extent here but I am going to look more carefully at the question of the distribution of stress and strain between the components of simple reinforced-concrete beams.

Treating the reinforced concrete as a unidirectional fibre composite, loaded in pure tension or compression in the direction of the fibres, the appropriate model to apply would be the parallel, homogeneous strain model, the starting assumption being that the strain in the concrete matrix (given the subscript 'm' for matrix in what follows) is the same as that in the steel (subscript 's'):

$$\varepsilon_m = \varepsilon_s$$

Then, for linear elastic behaviour, where $E = \sigma/\varepsilon$,

$$\frac{\sigma_m}{E_m} = \frac{\sigma_s}{E_s}$$

or

$$\sigma_s = \sigma_m \frac{E_s}{E_m} \quad (4.15)$$

Look at this relationship carefully. What it says is that the stress in the steel and the stress in the concrete are directly proportional to each other, and the constant of proportionality is the ratio of the Young's modulus values of the two materials.

SAQ 4.14 (Outcome 4.7)

Which parameters in Equation (4.15) does a designer have control over?

How and to what extent can they be varied?

Table 4.8 contains typical design data for the two materials. Using the values of Young's modulus given,

$$\sigma_s = 14\sigma_m$$

Table 4.8 Typical design data for reinforced concrete

	Strength MN m^{-2}	Design stress MN m^{-2}	Young's modulus GN m^{-2}	Strain to failure %
Concrete grade 40	26 (compression) 2.6 (tension)	17 (compression)	15 (long term)	0.35 (compression) 0.02 (tension)
Cold-worked mild steel	425 (0.2% tensile proof stress)	370	210	0.39

Although the steel experiences considerably greater stress than the cement, when the stress in the concrete reaches its tensile strength, $\sigma_m = 2.6$ MN m^{-2}, the stress in the steel will still only be $\sigma_s = 36$ MN m^{-2}. This is only about 8% of its proof stress and is well short of taking full advantage of the high strength of steel. The only way round this is *to let the concrete crack* so that the steel is subject to all the tensile forces. This is the basis of the design of reinforced-concrete beams and prevents us from using a simple model of composite behaviour like the ones we developed earlier, so a different approach is required.

First look at how a reinforced-concrete beam of straightforward geometry, such as that in Figure 4.38, behaves in bending. Note that since it is desired to benefit from its tensile strength, the steel should clearly be incorporated on the tensile side of the beam.

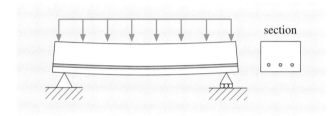

Figure 4.38 A simple rectangular beam in bending

As the loading on the beam is increased from zero up to failure, you can identify four different regions of behaviour. These are labelled I–IV on the load–deflection curve in Figure 4.39(a). In region I, both the steel and concrete behave elastically, the concrete in compression above the neutral axis and the steel and concrete in tension below it. This is reflected in the linear stress distribution through the thickness of the beam (Figure 4.39b), with a corresponding linear strain distribution.

Region I ends when the concrete on the tensile surface starts to fail at its fracture strain – for the grade 40 concrete in Table 4.8 this is about 2×10^{-4} (0.02%). In region II, the deformation of the beam is still elastic, but, with the cracked region increasingly spreading up to the neutral axis, the principal tensile resistance is provided by the steel; the bending stiffness of the beam is lower.

The onset of yielding in the steel signals the start of region III. The slope of the load–deflection curve decreases further, accentuated by the increasingly nonlinear compressive deformation of the concrete. The collapse of the beam in region IV, beyond the point of maximum load, is precipitated by failure of the concrete in compression at the upper surface of the beam.

Exercise 4.14

Describe the load–deformation behaviour if the beam were unloaded from each of regions I–III and then reloaded.

Thus, to make the most efficient use of the two materials, a beam should be designed so that the steel starts to yield before the concrete fails in compression. This is also a safe strategy since tensile yielding of steel is progressive and the beam remains load-bearing, while the crushing of concrete is catastrophic.

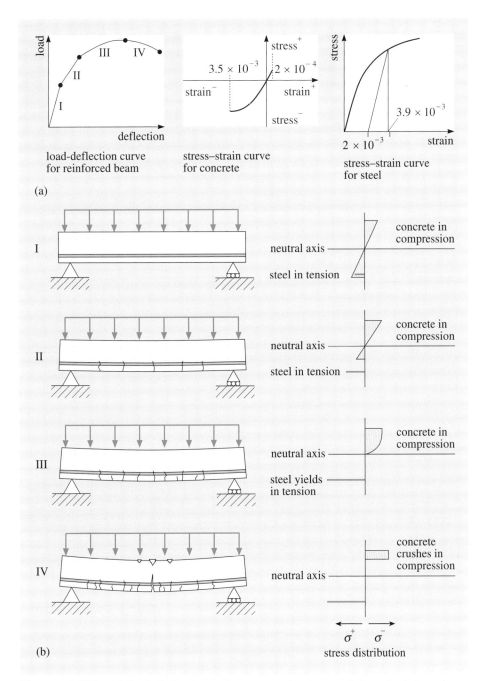

Figure 4.39 (a) load–deflection and stress–strain characteristics of reinforced concrete, concrete and steel, (b) loading to failure sequence and stress distributions for the beam. (Note, tensile stress is plotted to the left.)

7.2 A design example

I shall now look at an example of a reinforced-concrete beam of a size that you might well see in a building such as a multi-storey car park. It is 350 mm wide, 500 mm deep and has steel reinforcing bars of 32 mm diameter located 50 mm beneath its tensile surface. I shall assume that the concrete is grade 40 and use the data provided in Table 4.8. The table includes data for 'design stress' for both the steel and concrete which reflect the application of appropriate partial factors to the strength data using the limit-state design approach outlined in Part 1. Remember, these factors are semi-empirical values derived from a combination of statistical theories and real experience – we accept them and use them as directed by the relevant 'Codes of Practice'.

We want to know:

- the maximum bending moment that the beam can sustain;
- how many reinforcing bars are required.

The design strategy considers the beam to be loaded to region III of the load–deflection curve (Figure 4.39a), but to less than the maximum load. Thus, on the tensile side, the concrete has cracked and the steel has started to yield, while the concrete in compression has started to behave nonlinearly.

A partial factor also needs to be applied to the design loads – as this is only a worked example I shall use a factor of 1.5 and not specify the nature of the applied loads. The stress distribution for region III is shown in Figure 4.39(b), and I shall assume that the concrete under compression has reached the maximum design stress for grade 40, which is given in Table 4.8 as 17 MN m^{-2}. In region III the stress distribution is nonlinear; for mathematical convenience I shall approximate the nonlinear distribution to a constant compressive stress of 14 MN m^{-2} acting in the whole compressive region of the beam (i.e. above the neutral axis).

The combination of the distributed compressive stress in the concrete and the tensile stress in the steel results in an internal moment that is in equilibrium with the moment applied by the live loads on the beam and the dead load of the beam.

The compressive stresses above the neutral axis of the beam can be replaced by a single compressive force, C, acting at the centre of the compressive zone. This, together with the tensile force in the steel, T, forms a couple which gives the maximum internal moment which the beam can be designed to resist (Figure 4.40a). However, since the beam is no longer homogeneous throughout its thickness, it can no longer be assumed that the neutral axis lies in the centre of the beam's cross-section. Therefore I need to check the location of the neutral axis. This, together with expressions for the bending moment and cross-sectional area of reinforcing bars, is derived in my
▼ **Design calculations** ▲.

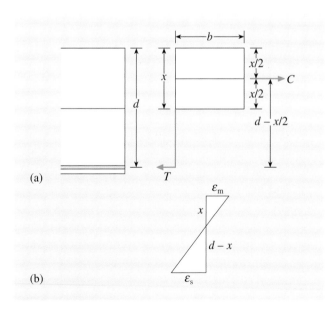

Figure 4.40 (a) internal moment of the beam; (b) assumed strain distribution in the beam (*cf.* Figure 4.39)

▼Design calculations▲

Neutral axis

Let the neutral axis lie a distance x beneath the top surface of the beam and the steel at a distance d (Figure 4.40a). The simplest way of approximately locating the neutral axis is to consider an equivalent solid material with a compressive strain, ε_m, in its upper surface (in concrete) and a tensile strain, ε_s, in its lower one (in steel). From the earlier analysis of beam bending we know that strain increases with distance from the neutral axis. In fact the strain varies *linearly* with the distance from the neutral axis. Thus we expect the linear, but not symmetrical, strain distribution shown in Figure 4.40(b).

Then, because the two triangles have the same angles (they are 'similar triangles'):

$$\frac{\varepsilon_m}{x} = \frac{\varepsilon_s}{(d-x)}$$

which can be rearranged to give:

$$\frac{x}{d} = \frac{\varepsilon_m}{\varepsilon_s + \varepsilon_m}$$

If we now assume that the amount of steel is such that it just starts to yield in tension as the concrete starts to fail in compression, using the failure strains given in Table 4.8, $\varepsilon_m = 0.35\%$ (in concrete) and $\varepsilon_s = 0.39\%$ (in steel). Substituting these values into the equation above gives the result:

$$x = 0.47d \qquad (4.16)$$

Cross-sectional area of steel, A_s

If we now consider the compressive forces in the concrete to be replaced by C, then horizontal equilibrium requires that:

$$C = T$$

Force is stress times area, so if the width of the beam is b this equilibrium can be rewritten as:

$$\sigma_m b x = \sigma_s A_s$$

Thus the cross-sectional area of steel required will be:

$$A_s = \frac{\sigma_m b x}{\sigma_s}$$

Substituting for x from Equation (4.16):

$$A_s = \frac{0.47 \sigma_m b d}{\sigma_s} \qquad (4.17)$$

Bending moment, M

Taking moments about the line of action of T gives:

$$M = C\left(d - \frac{x}{2}\right)$$

$$= \sigma_m bx\left(d - \frac{x}{2}\right) \qquad (4.18)$$

$$= 0.36\sigma_m bd^2$$

Based on a constant compressive stress in the concrete, $\sigma_m = 14$ MN m^{-2}, and given the dimensions of the beam, $b = 0.35$ m and $d = 0.45$ m, the maximum bending moment from Equation (4.18) will be:

$$M_{max} = 0.36\sigma_m bd^2$$
$$= 0.358 \text{ MN m} \qquad [\text{Note: } \sigma_m bd^2 = 0.99 \text{ MN m!}]$$
$$= 358 \text{ kN m}$$

Using the design stress for the steel, $\sigma_s = 370$ MN m^{-2}, the area of steel reinforcement from Equation (4.17) will be:

$$A_s = \frac{0.47\sigma_m bd}{\sigma_s}$$
$$= 2800 \text{ mm}^2$$

A 32 mm diameter bar has a cross-sectional area of 804 mm², so the A_s value would require four such bars.

Exercise 4.15

If the beam just considered ($M_{max} = 358$ kN m) spanned 5 m between its fixing points, estimate the maximum uniformly distributed load in kN m^{-1} that it can safely carry. Use a partial factor of 1.4 as appropriate to a dead load, and take $\rho_m = 2500$ kg m^{-3} and $\rho_s = 7860$ kg m^{-3}; maximum bending moment in a beam with fixed (or 'built in') ends of length L carrying a load w per unit length is:

$$M_{max} = \frac{wL^2}{12}$$

Hint: The maximum load the beam can carry will be its weight plus any applied load. The weight of the beam is obtained from its volume and density. Use the bending moment formula to establish a value for w.

SAQ 4.15 (Outcome 4.7)

Provide brief answers to the following questions:

(a) Why does the concrete have to crack for the reinforcement in a beam to be effective?

(b) Why is the steel on the lower side of the beam?

(c) Where is the neutral axis, and why isn't it at the centre of the beam?

(d) What defines the serviceability and ultimate loading limits for reinforced-concrete beams?

7.3 Prestressed concrete

One of the problems with reinforced-concrete members is that the concrete on the tension side, which is cracked and therefore doing no work, represents a considerable penalty in weight. In addition, vibration from traffic, for example, can cause repeated opening and closing of the cracks, leading to a gradual deterioration and crumbling of the concrete. Also, the cracks expose the reinforcing rods to possible corrosion. These problems can be overcome if the concrete can be subjected to an overall compression, so that none of it is in tension when the structure is loaded. This is known as *prestressing*.

Two techniques of prestressing are used: pre-tensioning and post-tensioning. In both cases, steel wires with a tensile strength of 1500 to 1800 MN m^{-2} are used to apply the compression to the concrete.

Pre-tensioning is used for the production, in a factory, of standard items such as beams and floor slabs. In this method the wires are pre-tensioned to about 1200 MN m^{-2} by hydraulic jacks and the concrete is cast around the wires. When the concrete has hardened sufficiently, the ends of the wire are cut. The elastically stretched wire is constrained from reverting to its unstressed length by its bond with the concrete. The contracting wire compresses the concrete and this inhibits cracking.

The stress distribution in a pre-tensioned beam under load is depicted in Figure 4.41.

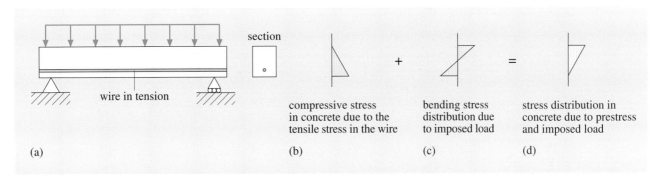

(a)   wire in tension

(b) compressive stress in concrete due to the tensile stress in the wire

(c) bending stress distribution due to imposed load

(d) stress distribution in concrete due to prestress and imposed load

Figure 4.41 Stress distribution in a pre-tensioned beam under load (*cf.* Figure 4.39)

When the pre-tensioned wire shown in Figure 4.41(a) is released, the beam is compressed. An indication of the resulting distribution of the compressive stress through the depth of the beam is sketched in Figure 4.41(b); it is greater at the bottom of the beam than at the top. Note the convention of indicating a compressive stress on the right of the vertical line. If a uniformly distributed external load is imposed as indicated in Figure 4.41(a) it will produce the familiar distribution of stress through the beam thickness indicated by Figure 4.41(c). The beam will remain in compression at all points on its depth (see Figure 4.41d), provided that the tensile stress due to the imposed load (the component to the left of the vertical line in Figure 4.41c) does not exceed the compressive stress created by the pre-tensioning.

Post-tensioning is used for on-site construction of large sections. In this technique large ducts are left when the structure is cast and wires are threaded loosely in the ducts. When the concrete has set, the wires are tensioned with hydraulic jacks acting on the end faces of the structure and the wires are wedged into the ends of the ducts (see Figure 4.42).

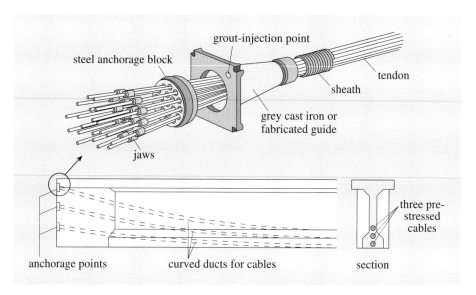

Figure 4.42 Post-tensioning

Liquid grout is injected under pressure to fill the ducts and once set this, together with the wedges, transfers the tensile stress in the wires to the concrete, thus putting it safely into compression. Post-tensioning can be used to assemble pre-cast units and in this way a long beam can be assembled on site from a number of short segments manufactured in a factory. This technique is used extensively in structures such as bridges and high-rise buildings.

Post-tensioning has an advantage over pre-tensioning in that the ducts can follow a curve. They can thus be placed to counteract the rather more complex distribution of tensile and shear forces that I alluded to earlier, whereas the wires used to pre-tension a beam must run in straight lines.

Compared with unstressed reinforced concrete, the use of pre-stressed concrete in, say, a bridge only requires about half the mass of concrete and a third the mass of steel, since the entire cross-section contributes to the bending stiffness. Thus, the potential economies are very large. Coupled with this are the improved aesthetic possibilities due to the more slender structural members.

One disadvantage that both unstressed and pre-stressed reinforced concrete share is the weight penalty of having to provide at least 50 mm of concrete between the steel and the outside world to protect the steel from corrosion. Thus there is considerable interest in developing ways of externally pre-stressing concrete, using high-strength materials that do not need protecting from the environment in the same way. Possibilities include carbon fibre, which is already in widespread use for the external repair of concrete structures, and other high-strength polymer fibres.

SAQ 4.16 (Outcome 4.8)

Describe briefly the distinction between pre-tensioning and post-tensioning techniques for reinforcing concrete. Why is steel with a yield stress of 1500 MN m^{-2} used as the reinforcement in pre-stressed concrete in preference to steel with a yield stress of 500 MN m^{-2}, which is used for conventional reinforcement (not pre-stressed)?

8 SUMMARY

The following are some of the important points covered in Part 4.

- Ashby charts, coupled with suitable merit indices, provide a 'first pass' comparison between different materials for given applications.

- The different profiles of properties exhibited by different materials make comparisons between them dependent on more than single values for given properties.

- Wood is a strongly anisotropic natural material that displays a wide range of properties in its various forms.

- Some control can be exerted over the properties of wood by converting timber into products such as plywood or the various types of fibre-boards.

- Concrete is a composite material that is strong in compression but weak in tension and is rarely used without some form of 'reinforcement'.

- The properties of concrete depend on the properties of the constituent cement and aggregate in proportion to their relative volume fractions in the mix.

- Steels for structural applications generally have a low alloy content providing low hardenability and high weldability.

- The properties of steels can be manipulated through the development of non-equilibrium microstructures, such as martensite, by heat treatment.

- Steel is incorporated into concrete to alleviate the latter's poor tensile strength. Further benefits can be gained by prestressing the concrete either by pre-tensioning or post-tensioning the reinforcing members.

This block has been about finding engineering solutions to the problem of building structures for shelter and protection. You have seen how such structures can be modelled as networks of members, and that these members have to stand up to both the load represented by their own weight and the applied loads of weather and human traffic. I have introduced you to the concepts of structural networks configured to sustain loads without movement, and I have shown you how dead loads and live loads are resisted by a structure by virtue of the internal forces that develop in its members. Finally, I have shown you a little of the materials engineering that goes into successful structural design. In that context, you have seen how it is possible to make choices between materials for structures.

The final component of Block 2 is presented in a slightly different fashion from the texts you have studied so far on this course. I should like you to see how the simple concepts you have learnt here (with some elaboration in certain cases) can be used to analyse successful structures from time past and to design new structures. The content for this part will be a series of readings taken from published sources, mostly on the Internet. You should turn next to the supplementary item *Block 2 Readings*.

9 LEARNING OUTCOMES

After you have studied Block 2 Part 4 you should be able to do the following.

4.1 Derive simple materials properties requirements from given outline structural designs. (SAQs 4.1 and 4.6)

4.2 Link characteristic tensile behaviour to particular materials and estimate values for specified tensile properties of a material from a stress–strain curve. (SAQs 4.2, 4.3 and 4.4)

4.3 Compare the potential performance of materials in a structural application, given appropriate data. (SAQs 4.4 and 4.7)

4.4 Explain the relationship between the orientation of the reinforcement in a composite material and the contribution of matrix and reinforcement to the modulus of the composite. (SAQs 4.5 and 4.11)

4.5 Apply simple concepts of merit indices to constrained problems in structural design and relate the outcomes to real-world solutions. (SAQs 4.8, 4.9 and 4.10)

4.6 Relate some aspects of the microstructure of steels to their mechanical properties with particular reference to planned and unintentional heat treatment. (SAQs 4.12 and 4.13)

4.7 Explain the principles of the design of reinforced-concrete beams and perform simple design calculations. (SAQs 4.14 and 4.15)

4.8 Understand and explain the principles behind prestressing of concrete for structural components. (SAQ 4.16)

ANSWERS TO EXERCISES

Exercise 4.1

By far the most likely failure mechanism will be a toppling of the tower when it bends, either due to instability in the mortar between the bricks or under wind loading. The bricks and mortar perform well under pure compression, but are very poor under bending stresses.

Exercise 4.2

Here is an account of the structures around me as I write. It is by no means inclusive or exhaustive.

I am sitting in a house made of brick with a tiled roof (although I can't see that from here). It has wooden door frames and doors. The roof I know to be supported on a frame consisting of wooden members. Outside I can see my greenhouse, which is quite different in design from the house. It consists of a metal frame with panes of glass providing the protective environment for the plants inside. I could even have considered my car parked outside with its metal body, or the computer casing on my desk made from metal and plastic. The list seems almost endless.

Exercise 4.3

I'll just consider the house and greenhouse discussed in my answer to Exercise 4.2:

- brick and tile – can take a heavy load, weather resistant, but heavy;
- wooden frames – reasonably strong, easy to cut and shape, may need treating;
- metal frame – strong, supplied in lengths, but may corrode;
- glass panes – transparent, weather resistant, but brittle.

Exercise 4.4

Pencil – squeezing doesn't have any discernible effect no matter how long I apply the pressure.

Eraser – shape changes under load, but returns to initial shape after load removed. The length of time I squeeze for doesn't seem to matter.

Poster adhesive – shape permanently changed after load removed. The deformation continues for some time under a constant load but it does seem to get 'stiffer' as time goes on.

Exercise 4.5

The most straightforward approach to this problem is to realize that the cross-sectional area of wire supporting the ceiling tile must remain the same if the stress in the wire is to be kept the same, regardless of the number of wires. If we call the diameter of each of the four wires d_4 and that of each of the three wires d_3, we can calculate the total areas for each scenario:

$$\text{total cross-section area of four wires} = 4 \times \frac{\pi d_4^2}{4} = \pi d_4^2$$

$$\text{total cross-section area of three wires} = 3 \times \frac{\pi d_3^2}{4}$$

Therefore,

$$\frac{3\pi d_3^2}{4} = \pi d_4^2$$

which can be rearranged to:

$$3d_3^2 = 4d_4^2$$

or

$$d_3^2 = \frac{4}{3}d_4^2$$

Therefore,

$$d_3 = \sqrt{\frac{4}{3}}d_4$$

Since d_4 is given as 1 mm, that means d_3 must be:

$$\sqrt{\frac{4}{3}} \times 1 \text{ mm} = 1.15 \text{ mm}$$

Exercise 4.6

(a) If the displacement is 0.80 mm over 25 mm, the shear strain is:

$$\gamma = \frac{0.80 \text{ mm}}{25 \text{ mm}} = 0.032$$

The shear force is:

$$F = A\tau = AG\gamma$$
$$= (0.025 \text{ m})^2 \times 2.8 \times 10^{10} \text{ N m}^{-2} \times 0.032$$
$$= 560 \text{ kN}$$

(b) The bulk modulus is:

$$K = -\frac{pV}{\Delta V}$$

So the change in pressure p to produce a volume change ($\Delta V/V$) of -0.01% is:

$$p = -\frac{K\Delta V}{V}$$
$$= -47 \times 10^9 \times (-10^{-4}) \text{ N m}^{-2}$$
$$= 4.7 \times 10^6 \text{ N m}^{-2}, \text{ or } 4.7 \text{ MN m}^{-2}$$

Exercise 4.7

Table 4.9 shows the data I derived from the two models, and these are plotted in Figure 4.43.

Table 4.9 Values of modulus for a composite over a range of compositions

Volume fraction filler	Modulus/GN m^{-2}	
	Homogeneous strain	Homogeneous stress
0	30.0	30.0
0.3	51.0	38.0
0.5	65.0	46.2
0.8	86.0	68.2
1.0	100.0	100.0

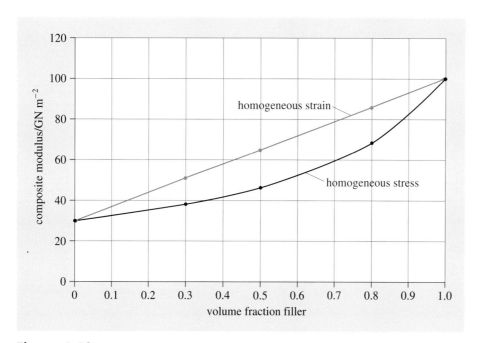

Figure 4.43 Composite modulus over a range of compositions

The key point that you should have observed is that the contribution that the filler makes to the composite modulus is more significant if the strain in both is equal and the stress is distributed between the matrix and filler. When the stress is equal and the deformation distributed, the filler becomes effective only at moderate to high volume fractions.

Exercise 4.8

The procedure is identical to that shown in the text until the point where an expression has to be found for *ab*.

Step 1

Using the same notation as in the text:

$$\frac{1}{mg} = \frac{1}{\rho gabL} \tag{4.19}$$

$$\frac{M}{I} = \frac{E}{R}$$

or

$$I = \frac{MR}{E}$$

$$ab^3 = \frac{12MR}{E}$$

or

$$ab = \frac{12MR}{Eb^2}$$

This goes back into Equation (4.19) to give:

$$\frac{1}{mg} = \frac{1}{\rho g L}\left(\frac{Eb^2}{12MR}\right)$$

and g can be eliminated to leave:

$$\frac{1}{m} = \frac{1}{\rho L}\left(\frac{Eb^2}{12MR}\right)$$

Step 2

$$\frac{1}{m} = \frac{E}{\rho}\left(\frac{b^2}{12MRL}\right)$$

Step 3

The merit index we want is E/ρ.

Exercise 4.9

Case 1

The merit index where the depth of the beam is fixed is E/ρ. A line with a slope parallel to the E/ρ line on the graph passing through steels also passes through titanium, aluminium and GFRP. Some of the ceramics and carbon-fibre reinforced plastics perform better.

Case 2

The merit index for a beam of variable depth is $E^{1/3}/\rho$. A line with this slope passes through the lower end of the high-density polyethylene (HDPE) area and the upper end of the low-density polyethylene (LDPE). Lots of plastics, wood, metals and ceramics perform better.

Exercise 4.10

(a) The appropriate merit index is $E^{1/3}/\rho$ which gives balsa, Sitka spruce and Norway spruce as the top three performers.

(b) Using E/ρ gives balsa, Sitka spruce and Scots pine.

These merit indices provide a measure of stiffness per unit mass. The mass of a floor joist is rarely a major concern in small house design. Stiffness per unit cost might be a better index to use. In addition we have the issue of design constraints that were mentioned earlier, where the size of the joist will generally be restricted within certain limits. So a balsa wood joist performs extremely well using the criteria, but the reality is that it would have to be unrealistically large in practice.

Exercise 4.11

The failure load of 30.6 tonnes corresponds to:

$$30.6 \times 1000 \times 9.81 \approx 3 \times 10^5 \text{ N}$$

Distributed over a cube face of 100 mm × 100 mm, this gives a compressive stress of approximately:

$$\frac{3 \times 10^5}{0.1 \times 0.1} \text{ N m}^{-2}$$

or roughly 30 MN m^{-2}. This is a Grade 30 concrete.

Exercise 4.12

(a) α iron is BCC ferrite; γ iron is FCC austenite.

(b) Just above the eutectoid temperature, only one phase, γ, is present, which is a solid solution of carbon in iron.

(c) Below the eutectoid temperature, two phases are present: α (ferrite) and Fe$_3$C (cementite).

(d) The amount of carbon that can be dissolved in the γ iron at 1100 K is about 1%.

(e) The maximum carbon content of the α phase just below the eutectoid temperature is about 0.05% (the right-hand limit of the α region in Figure 4.26)

(f) The only possible explanation for the whereabouts of the carbon is that it must have gone to the other phase, Fe$_3$C. You'll find out more about this phase by reading on in the main text.

Exercise 4.13

(a) The steel can transform to martensite or to a mixture of α ferrite and cementite (Fe$_3$C). (This appears as ferrite and pearlite in the microstructure.)

(b) The key factor in determining the likelihood of a particular transformation occurring is the cooling rate. If the rate is high enough to miss the nose of the TTT curve, martensite will form.

(c) It takes longer for the transformation from austenite to ferrite and pearlite to start in higher carbon content steels (the nose of the TTT curve is further to the right), so these are more likely to form martensite on cooling.

Exercise 4.14

Region I: Both components are still behaving elastically: removing the load and reloading would follow the original curve.

Region II: The beam would show partial elastic recovery on removing the load, the cracks in the concrete being closed by the tensile force in the steel. Reloading would be parallel to region II as the cracked beam would be less stiff than before.

Region III: The beam would have a permanent deformation on removing the load. Reloading would initially be parallel to region II.

Exercise 4.15

The maximum weight the beam can carry can be determined from the bending moment formula:

$$M_{max} = \frac{wL^2}{12}$$

M_{max} was established as 358 kN m. L is 5 m, so that gives:

$$w = \frac{12 M_{max}}{L^2}$$
$$= \frac{12 \times 358}{25} \text{ kN m}^{-1}$$
$$= 172 \text{ kN m}^{-1}$$

Applying the partial factor of 1.4 to this figure means that the maximum allowable loading on the beam is $172/1.4 \approx 123$ kN m^{-1}.

The weight of the beam per metre length w_b will be the sum of the weight of the concrete and the steel. Each will be obtained from the relevant cross-sectional area and the density of the material.

$$w_b = w_m + w_s$$

$$w_b = (A_m \rho_m + A_s \rho_s) g$$

Collecting the data for the beam we have:

$$A_s = 4 \times 804 \times 10^{-6} \text{ m}^2 \quad \text{(four bars at 804 mm}^2 \text{ each)}$$
$$= 3.2 \times 10^{-3} \text{ m}^2$$

The total cross-sectional area $= 0.35 \times 0.5$ m^2.

$$A_m = (0.35 \times 0.5) - A_s \text{ m}^2$$
$$= 0.17 \text{ m}^2$$

Therefore,

$$w_b = \left((0.17 \times 2500) + (3.2 \times 10^{-3} \times 7860)\right) \times 9.81 \text{ N m}^{-1}$$
$$\approx 4415 \text{ N m}^{-1}, \text{ or } 4.4 \text{ kN m}^{-1}$$

That suggests that the beam can carry an extra applied load of approximately $123 - 4.4 = 118$ kN m^{-1} (rounding down for safety).

ANSWERS TO SELF-ASSESSMENT QUESTIONS

SAQ 4.1

My summary of the requirements of materials in these two contexts is given in Table 4.10.

Table 4.10 Two possible types of structure to meet a design specification

Scheme	Brunelleschi's cupola	National Space Centre
Structure	hard shell	skeleton with flexible skin
Materials properties critical to providing structural integrity	compressive modulus and strength	appropriate strength and modulus to provide a rigid frame; flexibility for skin
Additional properties of importance to design	water resistance; durability	transparency; water resistance; durability

The umbrella-like Space Centre tower is a skeletal framework with a flexible skin. The umbrella does not normally need to allow light to pass through, but the requirement for water resistance is common to both. One extra aspect to the umbrella-frame design is the need for it to 'articulate' to allow it to be folded away when not needed.

The safety helmet is a hard-shell design but does not have to withstand constant heavy loads. The major materials requirement is one of toughness to withstand impact loading. (More about this in Block 5.)

SAQ 4.2

Most of these can be read from Figure 4.4(a), which is the graph for the steel specimen.

(a) The elastic limit is at about 950 MN m^{-2}.

(b) The tensile strength is about 1.15 GN m^{-2}.

(c) The 0.5% proof stress is about 1.05 GN m^{-2}.

(d) Since the reduction in area was 10%, the fracture stress was 10% higher than that shown in Figure 4.4, or about 1.1×1000 MN m^{-2} = 1.1 GN m^{-2}.

(e) As shown in Figure 4.4(b), the elongation at failure is the difference between the total strain at failure as shown on the stress–strain curve, and the recovered elastic strain. Using the plot in Figure 4.4(a), I made this to be about $(6.8 - 0.4)\% = 6.4\%$.

SAQ 4.3

(a) A is a strong, high-modulus material such as the steel described in Section 3.1. It has a high yield stress and shows some ductility.

(b) C is a lower-modulus, low yield stress material which is very much more ductile than A, and so is similar to lead.

(c) B is a fairly high-modulus material which shows no evidence of ductility or yielding. It fails at a very low strain within the elastic region, and so is typical of a brittle material such as glass, which fails by the growth of cracks.

(d) E has the lowest modulus by far. Its stress–strain curve is initially linear, then curves over to extend by a large amount for a very small increase in stress. At the end of this region, it becomes much stiffer, up to failure at several hundred per cent strain. This is the archetypal behaviour of materials like natural rubber, which are known as elastomers.

(e) D shows a low stiffness and nonlinear elasticity. It yields at a fairly low stress level and is also very ductile. This is typical of a plastic such as polypropylene.

SAQ 4.4

The most obvious point to note from the data is the dramatically lower modulus and strength offered by all the materials compared to steel. That means that, whatever else, structural members made from these materials would have to be much larger than the same component in steel to provide the same mechanical performance.

From the stress–strain curves we can say that glass fractures with no yielding, so is brittle and may therefore fail without much warning. Lead and polypropylene yield at fairly low stresses, which could be a problem. Rubber undergoes very large deflections at very low stresses, so is also unsuitable.

Altogether, steel looks like a pretty good choice.

SAQ 4.5

For the strain to be homogeneous between matrix and filler, any load applied to the composite must be applied directly to both components and they must be in intimate contact. This can only really happen with a fibre-reinforced composite with continuous fibres aligned in the direction of the applied load.

SAQ 4.6

(a) *Garden shed:* A few possibilities are feasible here. We could construct a shed based on a wooden frame with wood panelling and, say, a felt-covered wooden roof. Alternatively the shed could be constructed from stiff corrugated metal or plastic-composite sheeting.

(b) *Greenhouse:* Again there are alternatives. The greenhouse could be based on a stiff framework, probably metal, with glass panels to let the light through. Resistance to corrosion of the frame may also be a consideration. An alternative modern greenhouse may be based on a stiff frame but covered with a flexible, transparent, polyethylene.

SAQ 4.7

(a) A metal would be more suitable for the beam as it is strong both in tension and compression. The ceramic would not be suitable as the beam will be under tension on one side of its neutral axis and the brittle nature of ceramic would lead to failure under a very low load.

(b) The ceramic would be more suitable in an arch where 'wedges' of the material would be held (possibly under their own weight) in compression. The shape of the arch is critical to avoiding tensile stresses.

The metal may also be suitable, possibly with a wider range of design options which include tension as well as compression in the members.

SAQ 4.8

The key point is the extent to which the shape of the beam restricts or enhances the range of material options. If the only flexibility in design is the width of the beam, the highest-modulus materials are always going to dominate the choice. If, however, the depth can be varied, it is possible to utilize much lower-modulus materials without any loss in functionality.

Comment: Frankly, I'm not sure I've ever seen a polyethylene beam. There are clearly other factors at play that discourage that sort of thing. But, more to the point, it's rare that designers have a completely free rein over the dimensions of structures, and the extra depth required to compensate for the lower modulus of some materials may in reality be impracticable.

SAQ 4.9

I think the key factor at play is availability which, of course, is linked to cost. When my house was built, oak was quite widely available and oak trees still grew over large areas of southern England. Oak is now in short supply and hence expensive, whereas softwoods such as Sitka spruce are imported in large quantities from Europe and North America.

Oak did not perform well in terms of stiffness per unit mass because of its relatively high density – about twice that of Sitka spruce. If the two types of timber cost the same per unit volume (generally how bulk timber is costed) then oak will come out slightly better in terms of stiffness per unit cost. Oak would not have to be much more expensive relative to spruce to switch their relative positions in a strength/cost ranking.

SAQ 4.10

(a) By definition, the stress in the column at failure will be:

$$\sigma_p = \frac{F}{A}$$

(b) The cost of the column C will be the mass of the column multiplied by the cost per unit mass c. The mass of the column will be its volume times its density. The volume of the column is its area A times its length L. Thus,

$$C = AL\rho c$$

(c) The energy content Q will be the volume of the column multiplied by the energy content per unit volume of the material q. Thus,

$$Q = ALq$$

(d) By definition, the stiffness of the column is the force F per unit deflection δL. The deflection can be obtained from the standard formula for Young's modulus,

$$E = \frac{\sigma}{\varepsilon} = \frac{F/A}{L/\delta L}$$

This can be rearranged to give:

$$E = \frac{FL}{A\delta L}$$

$$\frac{F}{\delta L} = \frac{EA}{L}$$

(e) We can start with the expression from (b) above for the material cost:

$$C = AL\rho c$$

Since we want the minimum cost, we are actually looking for an expression for $1/C$:

$$\frac{1}{C} = \frac{1}{AL\rho c}$$

Now we need to bring in the expression from (a) above for the stress in the column at failure, $\sigma_p = F/A$. From that we get:

$$A = \frac{F}{\sigma_p}$$

Substituting this into the expression for cost, and separating the constants from the variables, gives:

$$\frac{1}{C} = \frac{1}{FL} \frac{\sigma_p}{\rho c}$$

So the merit index for the lowest-cost column of adequate strength will be $\sigma_p/\rho c$. The values of this merit index for each of the materials in Table 4.6 are given in Table 4.11 below.

So on the basis of these figures softwood has the least cost of these materials, with concrete a close second.

(f) The next exercise is similar to (e) but using the expression for energy content in (c) above:

$$Q = ALq$$

Substituting again for A in terms of force and strength and inverting the expression to give the *minimum* energy content gives:

$$\frac{1}{Q} = \frac{1}{FL} \frac{\sigma_p}{q}$$

The new merit index for the lowest energy content column of specified strength is σ_p/q. The values are again given in Table 4.11.

On this basis, softwood emerges as the best of these materials.

(g) Finally, we need to use the expressions from (b) above for the cost and from (d) above for the stiffness of the column. We are looking for the lowest cost per unit stiffness, which is to say the highest stiffness per unit cost:

$$\frac{\text{stiffness}}{\text{cost}} = \frac{EA/L}{AL\rho c}$$

Multiplying top and bottom by L/A gives:

$$\frac{E}{L^2 \rho c}$$

The length L is fixed, so the merit index of interest is $E/\rho c$. The data are again given in the table.

On this index, concrete emerges as the best-performing of the materials considered.

Table 4.11 Materials data and merit indices, from SAQ 4.10

Material	Relative cost per unit mass c/kg^{-1}	Density $\rho/\text{kg m}^{-3}$	Elastic modulus $E/\text{GN m}^{-2}$	Strength $\sigma_p/\text{MN m}^{-2}$	Energy cost $q/\text{GJ m}^{-3}$	Merit index for part (e) $\sigma_p(\rho c)^{-1}/\text{N m}$	Merit index for part (f) $\sigma_p(q)^{-1}/\text{N m kJ}^{-1}$	Merit index for part (g) $E(\rho c)^{-1}/\text{kN m}$
Cement	100	2400	30	16*	12	66.7	1.33	125
Concrete	50	2400	30	26*	8	217	3.25	250
Aluminium	2600	2800	71	200	500	27.5	0.40	9.75
Mild steel	600	7860	210	350	300	74	1.17	44.5
Softwood	500	400	11	50	3	250	16.7	55.0
Window glass	1000	2500	71	50	50	20.0	1.00	28.4
UPVC	1200	1450	3.0	50	70	28.7	0.71	1.72

SAQ 4.11

Concrete consists of a matrix of cement containing irregularly shaped particles of aggregate of various sizes. It certainly cannot behave according to the homogeneous strain model, as the load is transmitted to the aggregate via the matrix and the strain in the aggregate will therefore vary from that in the matrix. It follows that concrete is more likely to behave according to the homogeneous stress model.

SAQ 4.12

Several factors combine to cause concern about martensite forming at the root of the cantilever. First, martensite is brittle, so if the loading at the root exceeded the strength of the martensitic steel failure would be sudden and catastrophic. Second, both shear force and bending moment reach a maximum at the root of the cantilever (for both point loading and a uniformly distributed load) so this is where the structure is most highly loaded. Third, the sharpness of the joint can concentrate the stress and effectively amplify the load applied.

The normalized steel would be the preferred choice. Welding a quenched and tempered steel would 'undo' the heat treatment and dramatically alter its mechanical properties. You would not be able to predict with any certainty what the properties of the steel around the joint would be, and there is a high chance of martensite formation as the steel has been formulated to facilitate that during the quenching process.

These answers are hard to reach on your own, without someone to exchange ideas with. You should have no difficulty in following the argument, though.

SAQ 4.13

All three steels are designated as suitable for structural applications. The first has a yield stress greater than 355 MN m^{-2}, the second 275 MN m^{-2} and the third 410 MN m^{-2}. The first two steels are normalized, and the third is quenched and tempered. The minimum Charpy toughness of 27 J is achieved by the first at 0° C, by the second at –20° C and by the third at room temperature.

In short, the first is a medium-strength steel with a reasonable low-temperature toughness; the second is a lower-strength steel with superior low-temperature performance; the third offers a higher strength but at the expense of toughness at low temperatures.

The first steel might be a good general-purpose steel where structures are exposed to lower temperatures only on rare occasions. The second steel will offer advantages in colder climates, such as northern Europe and the upper latitudes in North America. Designs would have to be tailored to the lower strength of this steel. The third steel would be used where higher strengths are more important, for example in high-rise buildings, but only in relatively benign climates.

Both the first two steels could be welded since they were prepared in a normalized condition. The third could only be welded into a structure with extreme caution if unwanted changes in the mechanical behaviour of the material were to be avoided.

SAQ 4.14

(a) The stresses in each component will depend on the overall strain in the composite and the modulus of the component.

(b) A designer can adjust the strain resulting from a given applied load by altering the dimensions of the composite assembly or the volume fractions of each component.

(c) The Young's modulus of steel is effectively invariant.

(d) The modulus of concrete is obtained from the modulus values of each constituent of the concrete using the homogeneous stress (or series) model. In practice the modulus cannot be varied a great deal, as the amount of aggregate that can be incorporated in the concrete is limited to a narrow range defined by workability and the need to coat the aggregate particles adequately with cement.

SAQ 4.15

The modulus of the steel is much higher than that of the concrete but, at the same strain, the steel is experiencing only a fraction of its yield stress. The full benefit of the steel cannot be exploited until the strain in the composite has exceeded the failure strain of the concrete.

Beams naturally bend downwards under their own dead load and any applied load, inducing tensile stresses in the lower side of the beam. The steel is placed here to resist these tensile stresses.

The neutral axis of the beam is closer to the upper face of the beam. The beam effectively functions as a concrete strut above the neutral axis and a steel tie below. Because of the different stress–strain behaviour of the two materials, the strain distribution in the beam is asymmetrical, placing the neutral axis closer to the compression side.

The serviceability limit for a reinforced-concrete beam is reached when the steel exceeds its yield stress. An ultimate limit is reached when the concrete fails in compression above the neutral axis.

SAQ 4.16

In pre-tensioning, concrete is cast around the steel wires while they are under tension. The applied tension is released once the concrete has hardened and the bond between the steel and concrete ensures that the contracting steel puts the concrete into compression. In post-tensioning, the concrete is cast around ducts in which the steel wires are subsequently placed and tensioned. This tension acts on the end faces of the component, thus placing the concrete in compression.

The prestressed beam uses the steel in two ways: first as a provider of compression in the concrete, for which the steel is prestressed; second as a tensile load bearer. Thus it already starts part way up its stress–strain curve, so the higher yield material is preferred to give 'head room'.

REFERENCES

Ashby, M.F. (1999) *Materials Selection in Mechanical Design* (2nd edn) Butterworth Heinemann.

Benham, P.P., Crawford, R.J. and Armstrong C.G. (1996) *Mechanics of Engineering Materials* (2nd edn) Prentice Hall.

ACKNOWLEDGEMENTS

Grateful acknowledgement is made to the following sources for permission to reproduce material within Part 4 of this block.

Figures

Figure 4.2(a): © Edifice/Sayer; *Figure 4.2(b):* Courtesy of National Space Centre, Leicester; *Figures 4.10(a), 4.11 and 4.14:* Benham, P.P. *et al.* (1996) *Mechanics of Engineering Materials,* Prentice Hall, © Longman Group Limited 1996, reprinted by permission of Pearson Education Limited; *Figures 4.15 (a, b, c, d) and 4.19:* courtesy of Stan Hiller, Open University; *Figures 4.15 (e, f, g) and 4.35:* courtesy of Mike Levers, Open University; *Figure 4.16 (upper):* courtesy of Princes Risborough Laboratory Building Research Establishment, © Crown Copyright; *Figure 4.16 (lower):* Butterfield, B.G. and Meylar, B.A. *Three-dimensional Structure of Wood: An ultrastructural approach*, 1980 Chapman and Hall; *Figure 4.22:* Courtesy of Sheffield Newspapers Ltd; *Figure 4.30:* Reproduced with permission of The Principal and Fellows, Newnham College, Cambridge.

Every effort has been made to contact copyright owners. If any have been inadvertently overlooked, the publishers will be pleased to make the necessary arrangements at the first opportunity.

Course team acknowledgements

Part 4 of this block was prepared for the course team by Mark Endean, Alec Goodyear and Joe Rooney.

Trademarks

Blu-Tack® is a registered trademark of Bostik Findley Limited in the UK and Bostik Findley, Inc. in the USA.

Appendix: A world of materials

CONTENTS

1	Introduction	147
2	A materials classification system	153
3	Alloys	154
	3.1 Understanding alloys	154
	3.2 Ferrous alloys	158
	3.3 Non-ferrous alloys	162
4	Ceramics	165
	4.1 Glasses	165
	4.2 Domestic ceramics	166
	4.3 Engineering ceramics	166
	4.4 Natural ceramics	167
5	Polymers	168
	5.1 Thermoplastics	168
	5.2 Highly cross-linked thermosets	169
	5.3 Lightly cross-linked thermosets	169
	5.4 Natural polymers	169
6	Composite materials	170
7	Postscript	172
	Acknowledgements	173

1 INTRODUCTION

All engineers should possess some knowledge of the materials they use. Indeed, one would expect most engineers to be familiar with the make-up and typical properties of the major groups of engineering materials. However, it's not quite as simple as that. To start with, the wide range of engineering materials means that there can be many variations on each material depending on how it has been processed. So far we have principally considered materials just to be 'substance' that has a set of properties. We know from our common experience that these properties vary from material to material. Because they are such an integral part of everyday life, we all build up rule-of-thumb knowledge of the different substances that we encounter.

For instance, I don't suppose you would have any difficulty distinguishing between wood and metal. But how do you *know* they are different? The answer is quite simply that different substances *behave* differently when you do things to them. It's hard to imagine being able to cut treacle with a bread knife, for example, and you wouldn't normally try to saw up a sheet of glass as you would a sheet of hardboard or plastic.

Here, I will:

- show you how to *describe* different substances more precisely in terms of their physical characteristics and behaviour; and

- demonstrate to you some of the underlying *reasons* why particular substances behave in characteristic ways.

I shall concentrate on substances that are normally used in solid form and whose physical presence is essential for a structure to function properly. These are substances generally referred to in the engineering community as *materials*, and that is how I shall use the term from now on.

> ### Activity
>
> To get an idea of the enormous range of materials in common use, spend a few minutes looking at the structures around you, wherever you are reading this text, and see how many different materials you can recognize. You need not be specific about what a particular material is; just note it as similar to or different from the others.

I hope the activity has demonstrated the large range of structures around us and the different materials used in their construction. This raises the question of how the manufacturer in each case settled on one specific material, given such a wide variety to choose from? We can't answer the question for every material, but let's see if we can at least unravel a little of the complexity.

I have already suggested that materials differ from each other in that they behave differently under the same circumstances. Even if one material behaves only slightly differently from another, that's enough to identify it as different. We express this difference in behaviour in terms of a material's *properties*. From here we can start to develop a detailed description of the set of properties any material has in order to help us choose the right material for a job or to see why a material was suitable for, or perhaps not quite up to, the job it was supposed to do.

> ### Activity
>
> For the materials you identified in the previous activity, write down why they may have been selected for the particular task, and note any disadvantages you can see to their being used in that way.

Choosing the most suitable material for a particular job usually comes down to finding a material whose good points outweigh its bad points for the application. This does not just involve thinking about what happens to the structure in use; it also means taking into account the engineering that can be used to arrive at the right shape for the structure in the first place.

It is very unusual to find a single 'best' material for any application. The circumstances and available technologies have a strong influence on how you make things. Often the material that is the easiest to process does not have the best performance in use, and vice versa. Most choices of material represent a compromise between the properties you want and the ones you find acceptable.

Block 2 of the course examines the requirements of structures that have to bear loads, and explores the ways this can be achieved in terms of configuration and geometry. However, a sound structural design is not enough on its own; we must take care over the choice of materials for building the structure, as illustrated in Figure A.1.

Figure A.1 Good structural design and careful materials selection are necessary for protection against the elements

We shall concentrate just on describing the requirements for a material and the criteria for its selection. Even if we were to consider only materials that occur naturally in reasonable quantities on Earth, the possibilities and permutations available to the structural engineer would be enormous. Combining this with the possibility of using artificial 'engineered' materials can lead to quite a headache for the engineer in the role of a problem solver! Indeed, structural strength and stiffness, as discussed in Part 1 of this block, may not be the only considerations. Other issues may include resistance to weather, an ability to let through natural light, or aesthetic appeal, to name but a few.

So the configuration and dimensions of a structure will determine how the external loads are distributed as internal forces within the construction material. The integrity of the structure then comes down to the ability of the materials from which it is assembled to stand up to those forces, and the margin for error allowed in the design. Block 2 explores how the engineer goes about selecting materials, when faced with an almost infinite number of possibilities.

The purpose of this appendix is to provide you with a broad overview of the structural materials available to the engineer. We have seen that the properties of interest to the engineer vary from material to material, but we have not yet started to think about why. We are not going to go into this 'materials science' very far in this course, but what is useful is to realize that materials, and hence products, exist on a whole series of size scales. We are all familiar with the size range of tangible products, from a pin all the way up to a suspension bridge. We call variation on this scale *macrostructure*. Often, the macrostructure of a material can be seen with the naked eye – for example, as the grain in a piece of wood, or unwanted cavities (so-called *macroporosity*) in a metal casting or plastic moulding.

At the other extreme, you should also be familiar with the concept that the properties of materials are controlled by the type and arrangement of their individual atoms and molecules. Of course, this 'atom-level' structure is towards the lowest end of the size scale, and is very difficult to visualize directly. It is, though, possible to infer the structure of a material at the atomic scale from its physical and chemical behaviour. But such study belongs largely to the disciplines of Physics and Chemistry, and is not often pursued in depth in Engineering courses.

Much of Materials Science and Engineering is concerned with a size scale between the two extremes – too small to be seen with the naked eye, but much larger than individual atoms and molecules. This is the realm of *microstructure*.

The properties of solid materials are highly dependent on their microstructure. Furthermore, the microstructure of a given material can often be profoundly changed by how the material is processed. Thus, the properties of materials, and hence the structures that are made from them, depend on how they are processed. We shall return to these issues later in the course, but the important point to remember for now is that materials possess ▼**Structure at a variety of scales**▲ and it is a material's *microstructure* that often critically defines its properties and performance.

As you may remember from your previous studies, the properties required for a material to be easily processable very often conflict with those properties desired from the material in service. Many manufacturing processes use changes in microstructure to address this 'properties-versus-processing' dilemma. But now that we have an idea about what constitutes the microstructure of a material, we can specifically consider which materials are available to the engineer.

▼ Structure at a variety of scales ▲

Component geometry is an example of structure on a *macroscopic* scale. Look at Figure A.2, which shows a well-known bridge. The bridge has the structure it does because it was built to carry heavy vehicles safely across a river estuary. This structure was chosen, presumably, as a good balance between function and appearance at an acceptable cost.

Figure A.2 The suspension bridge crossing the estuary of the River Humber, in the north of England

If we look at the structure of a support cable for such a bridge (Figure A.3), we see that it is not a solid bar of material, but is 'woven' from many thinner strands of wire. This structure (still a macrostructure) is chosen for several reasons, including safety (the failure of one strand does not automatically lead to the failure of the whole cable) and some beneficial properties that cable structures have compared to large single strands, such as flexibility.

The structure story doesn't stop with the material for one strand, though. You may already know that steel is a mixture of iron and carbon. We shall see, later, that how the carbon affects the structure of the iron, on a microscopic scale, depends on the amount of carbon in the iron, and the heat treatment that the resulting alloy is given. Figure A.4 shows the *microstructure* of a typical steel, revealed by etching the surface with a dilute acid (note the scale bar on the photo).

Figure A.3 A steel cable from the Humber bridge

Figure A.4 Optical micrograph of steel (scale approximate)

This shows that as we look in closer detail, we begin to see that what we thought was just a plain metal surface has a lot of underlying structure to it. Once we've zoomed in so that we can see features as small as 10 μm (1 μm = 10^{-6} m), it becomes clear that the metal is composed of small individual 'grains' that are, in fact, individual crystals of the metal. This microstructure in turn determines the mechanical properties of the bulk steel, such as strength and toughness.

Microstructure can be changed by several means. Mechanical 'squashing' of the metal grains, either at ambient or elevated temperature, can be used to alter their shape. Alloying – mixing in trace quantities of other elements – can also be used to influence grain shape and crystalline composition, whilst heat treatment facilitates diffusion and microstructural equilibration. Mechanical processing, alloying and heat treatment are thus all tools that enable us to tailor the properties of the steel we make.

Figure A.5 zooms in still further, showing us more of the structure within the grains themselves. This transmission electron microscope picture shows the alternating laths of *cementite* (Fe_3C) and *ferrite* (Fe) that occur as an intimate mixture named *pearlite*. Influencing the structure at this level is more complicated, but it can be done, and again can help to tailor the material properties.

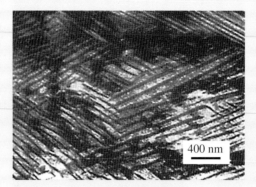

Figure A.5 Transmission electron microscope image of pearlite in 0.8%C steel (scale approximate)

Finally, we can zoom in to the scale of the atoms (Figure A.6). In this case, we're looking at atoms of carbon, one of the elements in steel, in a sample of pure graphite.

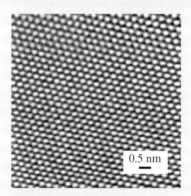

Figure A.6 Scanning tunnelling microscope image of carbon atoms in graphite (scale approximate)

The bonding between the atoms, and the structure they take up, critically influence the material properties; but there's little we can do to change it! Some materials are more useful than others because they have the right sort of atomic bonding, the right arrangement of atoms, and a microstructure that we can usefully manipulate. In Figure A.6, you should be able see that each carbon atom is surrounded by six others in a hexagonal pattern. (Remember that you experimented with hexagonal symmetry when building space-frame structures earlier in the block.) This is simply the way that carbon atoms arrange themselves in this structure (carbon is versatile in that it can adopt several atomic arrangements, as we will see later).

I shall refer to microstructure frequently in this section. It is a key factor in determining mechanical properties, and it can be greatly affected by the choice of manufacturing process for a material.

2 A MATERIALS CLASSIFICATION SYSTEM

What do you consider to be the main types of material?

Materials are most commonly divided into three groups:

- metals
- ceramics
- polymers.

I shall call these *classes* of material. In addition, I would add composites to the list. Composites are simply *physical* mixtures of more than one type of material, but the component materials can come from any of the other three classes. You may know that both structure and properties differ between the classes of material because of the strong links between chemical bonding, structure and properties. You may also know of differences in how these materials are made, processed and used in service.

Within each class of material there are many different examples. In such circumstances, it is often helpful to continue the idea of classification at various sub-levels within each major class to create what is termed a *hierarchy* of types. In the following sections, I shall briefly review each class of material to provide a framework for sub-classification. It is impossible to be comprehensive in this task, but you should be able to build on this framework as you learn more about materials during your engineering studies.

The start of a classification hierarchy of materials would look like Figure A.7.

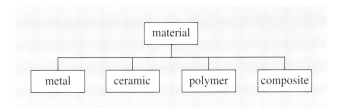

Figure A.7 A simple materials hierarchy

3 ALLOYS

Of the 95 naturally occurring chemical elements, all but about 20 are identifiable as metals. So it is perhaps not surprising to find that pure metallic elements display a wide range of properties. When mixed with each other and with some non-metallic elements they become even more versatile and form a large proportion of what are traditionally thought of as 'engineering materials'.

Apart from a few very esoteric exceptions, solid metals are crystalline, but the property that distinguishes the main engineering metals from the large number found in the periodic table is their ductility. Although high ductility confers both processability and toughness to a metal, most pure metals are too ductile to be of direct engineering use. However, by mixing metals together to form engineering alloys, and by controlling their microstructure, we can produce metal alloys with very different properties. For example, through a heat treatment both a high-strength aluminium alloy and a high-alloy steel (one with more than 5% alloying additions) can be converted from a soft, easily shaped state to the very strong but less ductile state in which they are used in service. These property changes are achieved by altering the *microstructure* of the metal, *not* by altering its overall composition.

> Ductility is a qualitative measure of how easily a metal can be deformed by stresses (such as those arising on drawing wire).

In the development of our understanding of metals from craft skill to science, some tools have evolved to help describe alloy systems and their behaviour. Perhaps the most important of these is the 'phase diagram', so next I am going to show you how these work, before moving on to look in more detail at some particular metals.

3.1 Understanding alloys

Although the term 'alloy' is frequently used for mixtures of many types of material – so that we have both polymer alloys and ceramic alloys – I am going to use it here in its original sense as a particular mixture of metals.

Most pure metals have high melting points and are relatively soft. If metals are mixed together, however, the result is an alloy that most often has a lower melting temperature than those of its pure constituents, and is also stronger than them.

This behaviour was exploited by Bronze Age peoples, when the dominant technology was one based on alloys of copper and tin. The addition of 8–12% tin to copper significantly reduces the melting temperature of the copper (making it easier to work with) and produces a stronger and tougher material.

So how do we work out what happens to the melting temperature of a metal as we alloy it with another metal? One way, of course, is simply to measure the temperature at which different mixtures of the metals melt, and then to plot this information on a graph. Such a graph forms the basis for what is called a *phase diagram*, which is a map describing the phases (solid, liquid or gas) present in a system over a range of compositions and temperatures.

A simple solid-to-liquid transition is a familiar occurrence: this is what happens to ice when it is taken from the freezer and warms up above its melting point. In alloy systems, however, the phases can be a lot more complicated. For one thing, there may be more than one type of 'solid' present. I shall come to this later. For simplicity, though, I am going to start with a *binary* alloy system – one that consists of a mixture of just two metals. The simplest type of binary phase diagram is shown in Figure A.8 and is typified by mixtures of copper and nickel (Cu–Ni).

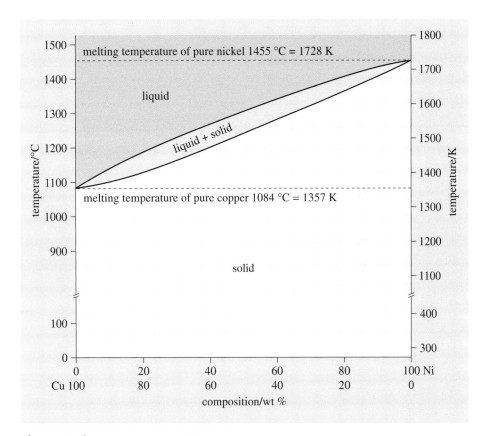

Figure A.8 The copper–nickel phase diagram

Copper and nickel show complete solid solubility, that is, they dissolve in one another as sugar does in tea, or gin in tonic. There is only one 'solid' phase, with varying compositions as shown by the horizontal axis. Because this region is a mixture of two elements and is solid, it is called a *solid solution*. So, overall, there are only two phases (liquid and solid) on our map, and we can use it to predict the melting temperature of any Cu–Ni alloy. If you look carefully you can see that for a given alloy composition there is a small temperature range where both liquid and solid exist at the same time. This semi-solid area is often termed the 'mushy' zone. Only the pure metals in the copper–nickel system have single, well-defined melting 'points'.

When looking at a phase diagram it is vital to understand that it represents a system *at equilibrium* at each temperature. Such a diagram is properly termed an 'equilibrium phase diagram'. But the conditions of perfect equilibrium in the solid state are rarely achieved in practice – a given mixture would have to be cooled almost infinitely slowly or held at a particular temperature for a very long time for all of the phases to come to equilibrium. Imagine putting a spoonful of salt in a glass of water and then putting the glass into a freezer without stirring. No doubt some of the salt will dissolve in the water before it freezes but, even then, the ice that forms nearest the bottom of the glass will have more salt dissolved in it than that further away. And it will stay very much the same for a very long time at a temperature of, say, –20 °C.

> A solid solution is like a solution of a solid in a liquid, or one liquid in another. The constituents are intimately mixed at an atomic level.

The complete solid solubility seen in the Cu–Ni system is rare. Restricted solid solubility is far more common and under these conditions the alloys concerned usually have lower melting points than either constituent metal. A *eutectic system* is one that has a distinct melting point at some composition that is lower than the final melting temperatures of all other compositions. The particular temperature is known as the *eutectic temperature* and the composition the *eutectic composition*. The lead–tin (Pb–Sn) system is an excellent example of such a eutectic system (although it is not much use as a structural material). I shall not go into much detail here, but you can see from its phase diagram (Figure A.9) that the eutectic composition is roughly 63% tin–37% lead (by weight), and the eutectic temperature is 183 °C. Notice that between about 20% tin and 98% tin, any alloy composition will begin to melt at this eutectic temperature, but melting will not be complete until a higher temperature is reached, and that higher temperature increases steadily towards the melting temperature of pure tin in one direction or pure lead in the other.

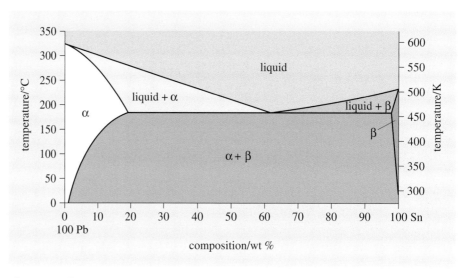

Figure A.9 The lead–tin phase diagram

There are now actually three phases on the diagram. For complicated alloy systems, the individual phases are generally assigned letters from the Greek alphabet, starting with α on the left-hand side and continuing alphabetically towards the right. On the lead–tin diagram the solid phases are α which is a solid solution of lead in tin, and β which is a solid solution of tin in lead. So at 50 °C it is possible to have an alloy of lead with up to about 3% tin dissolved in it. Adding any more tin will cause it to separate out as the β phase, just as there is only so much salt you can add to a glass of water before no more will dissolve in it!

Apart from these two solid phases, there is also the liquid phase on the diagram.

The large area in the centre of the diagram labelled α + β deserves careful attention.

Because at any particular temperature there is a limit to how much lead will dissolve in tin (α on the phase diagram) or tin in lead (β), we get a straightforward mixture of α and β over most of the range of composition. Note that this is *not* a mixture of bits of tin and bits of lead. Instead it consists of 'grains' of each of the two solid solutions, α and β. It is a

The first few letters of the Greek alphabet are:
α – alpha
β – beta
γ – gamma
δ – delta
The letter θ (theta) also appears on some phase diagrams to denote the formation of certain chemical compounds in the alloy.

two-phase solid solution and we refer to the size and distribution of the grains as the 'microstructure' of the alloy.

At a particular temperature, the difference between the microstructure of an alloy at one end of the range of composition and one at the other lies in the relative *size* and *number* of the two types of grain. Near the 'tin' end of the range where there is more tin than lead, you can expect to see more and bigger grains of α than of β. Conversely, at the 'lead' end, there will be more β than α. But it is important to realize that, at any given temperature, the proportion of lead to tin in the α phase will always be the same regardless of the overall composition of the alloy. The same goes for the β phase.

Looking at Figure A.9 you can also see that for most compositions the alloy solidifies over a wide range of temperatures. In contrast, for the eutectic composition, the alloy solidifies at a single temperature that is the lowest possible for the binary alloy system. The microstructure that forms at the eutectic composition is an intimate mixture of the two phases, as may be seen from Figure A.10.

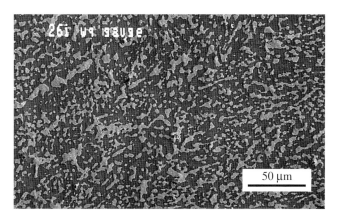

Figure A.10 Microstructure of the lead–tin alloy of eutectic composition, widely used in the electronics industry as solder (scale approximate). The α phase appears as the lighter areas surrounded by darker β

Nearly all modern alloys used for casting are eutectics. In practice, it is also possible to add small amounts of a third, or even fourth, element to the binary eutectic to enhance mechanical properties even further. It is also possible for one solid phase to change to other solid phases when it is cooled from an elevated temperature.

I am going to move on now to look at some specific types of metal to provide a few insights into how particular metals differ from each other, making them more or less well suited to different applications.

It is convenient to divide metallic materials into ferrous and non-ferrous alloys. Ferrous alloys are based on iron and range from plain carbon steels containing 98% iron to high-alloy steels containing up to 50% of alloying elements. All other metallic materials can be grouped together into the non-ferrous category which can then be subdivided into light alloys, heavy alloys, refractory alloys and precious metals. My complete classification of metals is shown in Figure A.11.

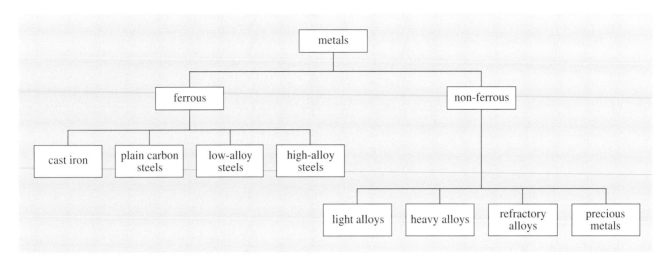

Figure A.11 A hierarchical classification of metals

3.2 Ferrous alloys

Ferrous alloys, and in particular steels, form about 90% of the total usage of metals in the world. The main reasons for this dominance are their relatively low cost and enormous versatility.

The versatility of steel owes a lot to ▼The wonder of allotropy▲ which allows steel to undergo two basic types of microstructural transformation depending on the speed with which it is cooled from high temperatures. But a simple change of crystal structure alone would not deliver the phenomenal microstructural control available to the steel metallurgist. What makes the difference is the manner in which the dissolved carbon is incorporated into the iron lattice. Steels dominate the field of structural metalwork.

▼The wonder of allotropy▲

In order to explain why steel can be processed to provide an almost bewildering array of property profiles, we have to look at the material at the atomic level. If the arrangement of atoms over a substantial number of atoms within a solid is regular, it is considered to be a *crystalline* solid. What this means is that the local environment of every atom in the solid is the same, just as the environment of each brick in a long tall wall is the same – that is, apart from at the edges, any brick is surrounded by the same pattern of other bricks. One difference between a crystalline solid and any wall, though, is that the atoms in a crystal go on and on in all three dimensions for millions of atoms before the edge of the crystal is reached. Virtually all metals are crystalline, but the exact geometric way in which atoms are packed to form the crystal may vary from metal to metal.

Some solids even have different crystal forms at different temperatures. The existence of several different forms of the same element we call *allotropy*. Many elements are allotropic. A familiar example is carbon, which can exist as both graphite and diamond – heating a diamond above about 600 °C can trigger a change back to graphite, the stable form of carbon at room temperature (and make you very unpopular). But the allotropy of iron is especially important as it is critical to understanding why steels are so useful.

At room temperature, pure iron exists in the particular crystal form shown in Figure A.12(a). On heating up, it undergoes an abrupt change to the form shown in (b). This happens at precisely 910 °C. If heating continues, then at 1390 °C the iron suddenly reverts to structure (a). The diagrams show what is known as the *unit cell* of the crystal, which is the basic building block for the whole crystal. Within one crystal of pure iron there will be many millions of these unit cells. The structure in (a) is referred to as a *body-centred cubic* (BCC) structure as there is an atom at the centre of each cube of atoms. The structure in (b) is a *face-centred cubic* (FCC) crystal, as there is an atom at the centre of each face of the cube.

Any crystal structure comes about through complicated interactions between thermal energy and bond energy. Complete explanations involve the tricky subject of thermodynamics. For our purposes it is enough to know what happens without knowing why.

(a)

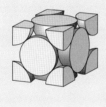

(b)

Figure A.12 Typical crystal structures: (a) body-centred cubic unit cell, (b) face-centred cubic unit cell

Plain carbon steels are the cheapest and most commonly used steels and despite their name can contain deliberate additions of up to 1.65% manganese and 0.6% silicon. The strength of plain carbon steels is primarily a function of their carbon content, as may be seen from Figure A.13. Unfortunately, the ductility of these steels decreases as the carbon content increases, and they respond only rather weakly to heat treatment. In addition, plain carbon steels lose most of their strength at elevated temperatures, can be brittle at low temperatures, and are susceptible to corrosion in most environments! Their big advantage is that they are relatively cheap, costing typically £168 per tonne at 2003 prices.

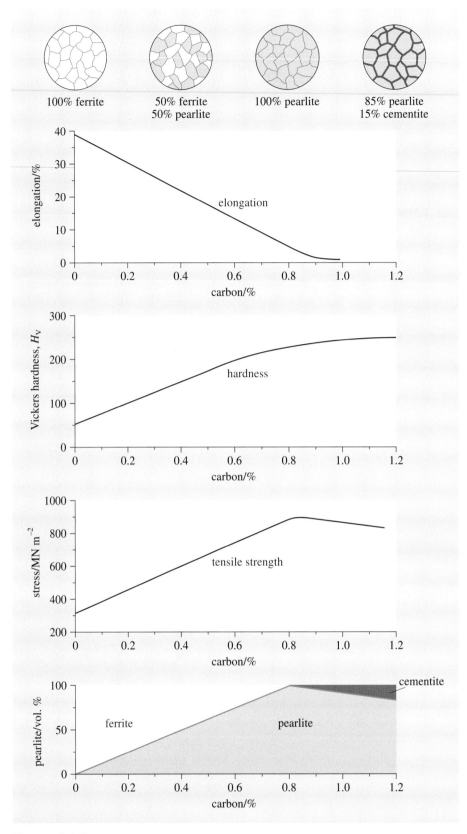

Figure A.13 The variation of mechanical properties with carbon content in plain carbon steels

The limitations of plain carbon steels can be addressed by adding other alloying elements. The commonly used additives and their effects on steel properties are shown in Table A.1.

Table A.1 Benefits of major steel alloying elements

Element	Amount added (wt %)	Effect
aluminium	< 0.1	deoxidizes melt, restricts grain growth, increases surface hardening on nitriding
boron	0.001–0.003	increases hardenability
chromium	0.5–4.0	increases hardenability and strength
	4.0–18	increases resistance to corrosion and oxidation
cobalt	1–12	improves cutting-tool life at high temperatures, increases hardness
manganese	0.25–0.4	mops up sulphur by forming MnS inclusions, hence prevents brittleness
	> 1	improves hardenability, strength and hardness
molybdenum	0.2–5	improves hardenability, strength and hardness, enhances creep strength
nickel	2.0–5	increases toughness particularly at low temperatures
	12–20	improves resistance to corrosion and oxidation
silicon	0.2–0.7	deoxidizes melt, increases strength
	2.0	improves elastic properties (e.g. for springs)
	> 2.0	decreases a.c. magnetization losses (transformer cores) Increases electrical resistivity
sulphur	0.08–0.15	when combined with Mn forms MnS inclusions which aid machinability
tungsten	up to 20	increases hardness at room and elevated temperatures, improves hardenability
vanadium	up to 5	increases strength whilst retaining ductility at room and elevated temperatures

Low-alloy steels contain up to about 5% of combined alloying elements. You have to pay for the improved properties conferred by the alloying elements and the increased quality control needed in manufacture, and low-alloy steels cost typically 20% more than plain carbon steels. Alloying elements vary both in price and availability but, by adjusting the proportions of different additives, steels can be produced that have almost identical mechanical properties but different chemical compositions. Alloying elements can also be strategically important if production is restricted to politically unstable countries. Whilst this may be of concern at a national level, for most users such factors are already built into the price. Thus, in most cases the best steel to use is the cheapest one that possesses the desired properties.

Stainless steels, as their name suggests, are used principally for their corrosion resistance. This ability to resist attack is due to a self-healing and non-porous chromium oxide film that forms in the presence of oxygen. A minimum of 12% chromium is required to form this film. In the construction industry, stainless steels are increasingly used where corrosion resistance is important, such as in the repair and restoration of historic buildings. Stainless steels are subdivided according to their microstructure as austenitic, ferritic and martensitic – terms that you will meet again when we look at structural steels more closely.

Tool steels are high-alloy, high-carbon steels that are used to form or cut other materials. Thus, a tool steel must have excellent wear resistance, high hardness at elevated temperatures and high toughness. There are two main tool-steel types depending on whether molybdenum (M-type) or tungsten (T-type) is the major alloying element. All high-alloy steels are expensive, and typically cost five to ten times as much as plain high-carbon steels.

Cast irons are the final class of ferrous metals. The metallurgy of cast iron is horrendously complicated and a comprehensive treatment is well beyond the scope of this course. However, all cast irons are based on alloys of iron, carbon and silicon. They contain more carbon than steels and thus their microstructure can contain graphite (essentially pure carbon) as well as the cementite (Fe_3C) usually found in steels They are sub-classified according to the form of the carbon-containing phases. Microstructures of typical cast irons are shown in Figure A.14.

3.3 Non-ferrous alloys

Non-ferrous alloys are normally more expensive than ferrous alloys. They are, therefore, most often used when specific properties, superior to those of ferrous alloys, are sufficiently important to justify the higher cost. Particular non-ferrous alloys can be chosen for their excellent corrosion resistance, electrical or thermal conductivity, specific strength (strength per unit mass or volume) or processability.

It is convenient to divide non-ferrous alloys into four major groups – light alloys, heavy alloys, refractory metals and precious metals.

Light alloys have low densities and can be heat-treated, after processing to shape, to develop high strengths. They are also tough, and so provide excellent strength and stiffness per unit weight. These alloys, as may be predicted from this brief property profile, are heavily utilized by the aerospace industry.

Aluminium alloys are probably the most economically important of the non-ferrous alloys, and find widespread use in the construction industry, where low weight and good corrosion resistance are beneficial. Aluminium has been highly developed, so that alloys are available which optimize conductivity, formability or specific strength, and these are used where specific properties are major requirements.

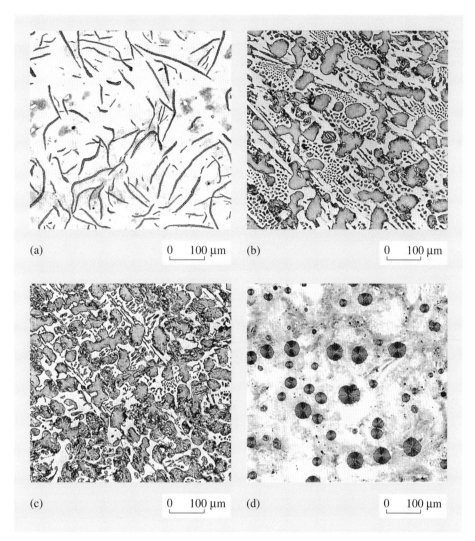

Figure A.14 Typical cast-iron microstructures: (a) grey iron, (b) white cast iron, (c) malleable cast iron, (d) nodular cast iron

Magnesium is the lightest metal in general engineering use. Although magnesium alloys are weaker and more expensive than aluminium alloys, their low density makes them competitive for a number of transportation uses, such as high-performance cast wheels and engine crankcases.

Titanium is allotropic like iron and, as with iron, the addition of alloying elements and thermochemical treatment gives a wide variety of properties. This flexibility is combined with a low density, so that titanium alloys exhibit the highest specific strengths found in engineering alloys. What is more, titanium alloys still possess good strengths at very high temperatures.

Heavy alloys are not, in fact, all that heavy, having densities typically just greater than steel and less than half that of refractory or precious metals (see below). The adjective 'heavy' has come into common use simply to distinguish these engineering metals from the light alloys. They may also be distinguished from the other two categories by their use and, in fact, include a large number of familiar alloys such as solders (lead and tin, silver and tin), brasses (copper and zinc) and bronzes (copper and tin).

Refractory metals are characterized as having melting points above 1600 °C. The most important metals in this group are tungsten, molybdenum, tantalum and niobium. Apart from their high-temperature uses (Figure A.15) their main use is as alloying elements for steels.

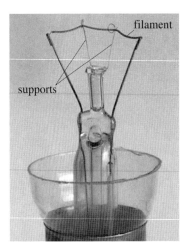

Figure A.15 A tungsten light-bulb filament and its support wires made from molybdenum

Precious metals form the final group of non-ferrous metals, and include gold, silver and the platinum-group metals. Despite their high cost, most precious metals are used for more than jewellery and financial investment – the main use of silver is in photographic emulsions, for example. They have considerable use due to their particular chemical properties. Thus gold is extensively used in the electronics industry for electrical contact, where its corrosion resistance is paramount, and the platinum-group elements find widespread use as catalysts, not least in automobile exhaust catalytic converters.

4 CERAMICS

Ceramics are mostly compounds formed between metallic and non-metallic elements, although you will find a number of materials classed as ceramics here, for convenience, which do not fit this description. The non-metallic element is often oxygen, as in magnesia (MgO) or alumina (Al_2O_3), but can also be carbon (silicon carbide for instance), phosphorus, nitrogen or boron (as in boron nitride). This definition of ceramics covers a wide variety of materials as there is obviously a large number of possible combinations of metallic and non-metallic elements. Examples of useful ceramic materials include brick, stone, porcelain and glass, in addition to the refractory compounds mentioned above.

The microstructures of ceramic materials are usually of the following three types:

1. fully crystalline
2. fully amorphous (glasses)
3. partially crystalline ceramics where crystalline grains are cemented together by an amorphous matrix, the latter usually being the weaker of the two phases.

The high melting points and lack of ductility of ceramics severely limit the ways in which they can be processed. The production of many ceramic products, therefore, involves the compaction or consolidation of the materials in powder form. The inevitable result of this is that it is very difficult to manufacture products that are entirely free of pores, and pore-like defects in materials generally degrade their mechanical properties.

4.1 Glasses

Glasses are used in large quantities, their annual consumption being close to that of aluminium. Many recent architectural developments have exploited glass extensively to create striking buildings.

Most glasses are based on silica (SiO_2) which, you will note, is not a compound of a metal. Pure silica is, in effect, a polymer consisting of chains of linked SiO_4 tetrahedra (the oxygen atoms being shared with the adjacent tetrahedra), and has a high softening temperature, which restricts its use in practice to a few specialized applications. However, the addition of compounds (often called fluxes) such as CaO, Na_2O and B_2O_3 enables a lower processing temperature to be used. Perhaps the two most important glasses of this type are *soda-lime glasses*, containing sodium and calcium oxides to reduce the melting temperature and decrease the crystallinity, and *borosilicate glasses*, where all of the lime and much of the soda is replaced by boric oxide (B_2O_3) to produce a glass with a low coefficient of thermal expansion.

Glasses are intrinsically very brittle, but useful improvements in toughness can be obtained if the glass is engineered to have a residual compressive stress at its surface: an applied tensile stress then has to exceed the residual compressive stress before any crack-like defects in the surface can begin to propagate through the material. There are two principal ways of achieving this:

1. Physical treatments involve cooling the glass very rapidly from just below its softening point to induce compressive stresses in the surface layers. Some automobile windscreens are 'thermally' toughened by this method.

2   Surface compressive stresses can also be introduced by chemical methods. These involve chemically altering the composition of the surface of a high-expansion glass to that of a low-expansion glass, by introducing other ions. Chemical methods are more expensive than their physical equivalents and so are used only for specialized applications such as aircraft windows and safety glasses.

4.2 Domestic ceramics

I have chosen the term 'domestic ceramics' to describe those ceramic materials that are familiar in everyday objects – pottery, bricks and cement. All except cement are based on clays, of which the most common example is kaolinite $[Al_2Si_2O_5(OH)_4]_2$, which in the unfired condition is soft and highly plastic.

After shaping this soft plastic clay into some desired shape, the water must be dried off, often much of it at room temperature, and then the components can be 'fired' (or 'sintered') at a high temperature. The composition of the clay and the temperature to which it is heated determine the resulting material. *Earthenware*, the material used for flower pots, house bricks and roof tiles, results from solid-state consolidation of the clay particles into a highly porous material. If the temperature becomes high enough to cause a liquid phase to form and flow around the particles, the end product is known as *stoneware*. *Porcelain* is obtained by using a different combination of starting materials to give a microstructure which is, in effect, a fibre-reinforced composite (Section 6) that is both strong and tough.

A further development of porcelain gives *vitreous china*, which has the same ingredients as porcelain, but in different proportions. This is used for run-of-the-mill domestic crockery and sanitary ware, and has a fracture stress 50% higher than porcelain, but without the same translucency and quality of appearance.

Classifying *cements* as domestic ceramics simply highlights one of the limitations of hierarchical structures. They share neither the composition nor the consolidation mechanism of the other materials in the group, but it is convenient to put them here. Cements are consolidated by chemical reaction with water (hydration) rather than by firing, and they consist of crystalline particles of lime (CaO), silica and alumina held together by a silicate gel. Cements are examined in more detail in Part 4 of this block.

4.3 Engineering ceramics

Engineering ceramics have, as their name suggests, improved mechanical properties when compared to more traditional ceramics and are designed for more stringent engineering use. However, this invariably makes them more expensive.

Alumina is perhaps the most widely used ceramic material not based on silicon. Aluminas are used for high-quality electrical and thermal insulation, abrasives and cutting tools.

Carbides are another important class of engineering ceramics. Tungsten and titanium carbides are both used for wear-resistant products, such as cutting tools and metalworking dies, whilst silicon carbide is also used for engine and turbine parts. It could possibly be argued that graphite and diamond, which are both forms of pure carbon, are really polymers. However, their properties are so unlike those of the materials we commonly think of as polymers, and so similar to many other ceramic materials, that a classification devised for selection between materials will always include them in this latter class. Both have wide engineering uses, diamond as an abrasive and graphite as a dry lubricant, and increasing use is being made of carbon fibre. In this material the graphite-type sheets of carbon atoms form into concentric tubes, producing an extremely strong but anisotropic material. This is now used in a variety of relatively mundane goods, from golf club shafts to umbrellas, as well as in performance-critical areas such as aerospace.

Nitrides include cubic boron nitride (CBN) and silicon nitride. The former is used as an abrasive and to make imitation diamond rings, whilst the high creep resistance of silicon nitride has enabled it to become a candidate material for many structural applications in gas turbine and automobile engines.

Borides offer excellent strength, hardness and wear properties, but on their own are very brittle. However, titanium boride is included as reinforcement into metal matrices to form metal matrix composites (MMC). These show physical characteristics of both ingredients (high strength/hardness and wear resistance from the reinforcement and ductility from the metal matrix). MMCs are used in automotive and aerospace industries.

4.4 Natural ceramics

Natural materials are either geological or biological in origin. Biological systems, such as the human body, generally produce a mixture of polymeric and ceramic materials, mostly containing carbon in some form and many of which are composites. The products of geological processes, however, are exclusively ceramics, with silicon being the predominant element. The study of natural ceramic materials, which include all the various types of building stone, is a huge subject.

5 POLYMERS

Most useful polymeric materials are compounds of carbon. It is possible to divide all polymers into two major classes – thermoplastics and thermosets.

Thermoplastics soften on heating and harden on cooling, irrespective of how many times this process is repeated. It is useful to subdivide thermoplastics according to whether they can crystallize or not.

Thermosets, on the other hand, form primary bonds between molecules during processing, commonly called *cross-links*. Once these cross-links have formed, the thermoset will not melt without degradation. Thermosets can be further subdivided according to the extent of cross-linking between the chains.

As mentioned above, many natural materials are polymeric in nature. With the obvious exception of natural rubber, most are, however, not processed in a way which alters their microstructure, so their classification as either thermoplastic or thermosetting is not very helpful.

For the purposes of Block 2, I do not need to develop quite the same level of detail concerning how polymers behave as I did for metals. I shall, therefore, just introduce some of the types of polymer that are commonly used in the construction industry.

5.1 Thermoplastics

Polyalkenes form the most widely used group of thermoplastics. Polyethylene and polypropylene are the main members. One of the curiosities of the polymerization process for polyethylene is that the molecule can develop branches, and these influence the degree of crystallization possible in the solid polymer. The amount of such branching depends on the polymerization technique adopted, and the end product is classified according to the degree of crystallinity, expressed as its density. Blends of the different densities are possible, leading to low-, medium- or high-density polyethylenes (LDPE, MDPE and HDPE, respectively).

Polyamides (nylons) have the characteristic of forming hydrogen bonds between molecules which confers on them a high degree of crystallinity. The major disadvantage of hydrogen bonding is that it brings with it an affinity for water. Nylons therefore absorb moisture, with an accompanying dimensional change and reduction in stiffness, but increase in toughness, as the water reduces the degree of bonding between the polyamide molecules.

Polystyrenes are second only to polyethylenes in volume of use. They are typically amorphous and thus, although they have reasonable tensile strengths, they have low toughness and are susceptible to creep. As a result they are widely used in modified forms, as foamed plastics in insulation, and incorporating rubber particles to provide an energy-absorbing mechanism during impact. This latter modification yields polymers such as high-impact polystyrene (HIPS) and acrylonitrile-butadiene-styrene (ABS).

Poly(vinyl chloride), **PVC**, has a unique set of characteristics. In its unmodified form (commonly known as uPVC or PVCu, the 'u' standing for 'unplasticized') it behaves as a classic amorphous thermoplastic and is extensively used for doors and window frames in place of wood. It has the ability, however, of absorbing high volumes of low molecular weight organic compounds, which then have the effect of 'plasticizing' or 'flexibilizing' the polymer. One of the biggest uses for plasticized PVC is in insulation for electrical wiring.

5.2 Highly cross-linked thermosets

The strong intermolecular covalent bonding present in thermosetting plastics makes them harder and more brittle than thermoplastics, but gives them greater thermal stability and creep resistance. The low toughness values of all these polymers mean that it is extremely unusual to come across them in an unmodified state, and they are commonly combined with fibrous or particulate materials ranging from wood to glass.

Epoxies are perhaps the most familiar thermosets owing to their use in adhesives such as Araldite®. They are also widely used as matrix materials for fibre-reinforced plastics. They are, however, relatively expensive.

Polyesters are extensively used as a lower-cost alternative to epoxies in composites.

Phenoplasts are the oldest class of thermosets, and are based on phenol-formaldehyde. Other thermosetting plastics involving the use of formaldehyde as one of the reacting species are urea-formaldehyde, which is typically used for electrical fittings, and melamine-formaldehyde, which is used for tableware and as the matrix material in decorative laminates for domestic work surfaces. Because of their nitrogen content, they tend to be grouped together as aminoplasts.

5.3 Lightly cross-linked thermosets

All the lightly cross-linked thermosets are rubbery in normal use. Rubbers are very commonly combined with other materials either for reinforcement or cost reduction, and the most common reinforcing additive is finely divided carbon powder, known as 'carbon black'.

Natural rubber is the rubber used in the greatest volume and is traded on the commodity markets. Combined with carbon black and other fillers, it provides a range of rubber compounds of varying hardness.

Synthetic rubbers with similar compositions to natural rubber come in a wide range of chemical variants including styrene butadiene rubber (SBR), polychloroprene (or neoprene) and nitrile rubber.

5.4 Natural polymers

The interesting point to note about natural polymers is that during processing into different shapes, great pains are taken to preserve, or even enhance, the underlying microstructure of the material. Even extreme examples of processing such as the manufacture of paper do not destroy the fibrous character of the material, but rather rearrange the various constituent fibres to create more of an isotropic structure than was the case in the original material. The majority of natural polymers are *fibres*. This includes materials such as wool and cotton that are used as discrete fibres or collections of fibres, and also biological fibres, such as ligaments and tendons, which are primarily composed of the fibre collagen.

6 COMPOSITE MATERIALS

A composite material has two or more components, which combine to produce properties that cannot be achieved by any of the components individually. This definition actually encompasses over 90% of engineering materials, since multiphase materials are the norm rather than the exception. However, materials such as steels, aluminium alloys and polyalkenes are not normally considered to be composites, since the components of these mixtures are the result of segregation of constituents and separation of phases of a material that is homogeneous in the fluid state. The composites classified as such here are mixtures of materials that remain discretely separate throughout the cycle of processing.

In any composite, one of the phases present is continuous and forms the matrix. The function of the other phase is then to reinforce the matrix. The properties of the resulting composite are largely controlled by the size and distribution of the reinforcing phase.

There are three principal ways of reinforcing a material.

Particulate composites contain dispersed phases that are much larger than those found in dispersion composites. The particles are larger than 10 µm and their volume fraction is usually greater than 20%. Strengthening of particulate composites is mostly due to restriction of matrix deformation and is critically dependent on the relative elastic properties of the matrix and reinforcing phase. The relatively poor properties of bulk ceramics have ensured the development of a large number of composites containing ceramic reinforcement. Two well-known examples of ceramic particulate reinforcement are in grinding wheels, where the hard cutting particles are embedded in a vitreous matrix, and *cermets*, in which particles of hard ceramic are bonded together with metal. One important cermet consists of tungsten carbide in cobalt, which is extensively used for cutting tools. Many glass-filled and most mineral-filled polymers can also be considered as particulate composites.

Fibrous composites encompass a wide variety of materials where a matrix is used to bind together fibres and to protect their surfaces from chemical attack. Probably the best known examples of this type of composite are the fibre-reinforced plastics (FRP), which range from the glass fibres in a polyester matrix used for minor repairs to vehicle bodywork, to the carbon-fibre-reinforced epoxy (known as CFRP) used in aerospace and other high-specification applications. Steel-reinforced concrete is another example that illustrates the importance of the matrix as a surface protection against corrosion. This particular example is studied in Part 4 of this block. Recently, significant effort has been expended on producing ceramic-fibre-reinforced metal matrix composites. Typical fibres are silicon carbide or alumina and, owing to their improved stiffness, high-temperature and wear properties, such fibre-reinforced metals (FRM) will undoubtedly become more popular in the future. *Natural composites* such as wood and leather are almost exclusively polymer matrix fibre composites.

Dispersion composites are characterized by microstructures consisting of 1 to 15% of fine particles, ranging from 10^{-5} mm to 2.5×10^{-4} mm in diameter. The fine particles introduced harden the material by impeding dislocation motion in much the same way as precipitation hardening induced by heat treatment. Their principal advantage over precipitation-hardened materials becomes apparent at high temperatures, when precipitates coarsen easily. This is because, being physically mixed, the fine particles can have melting points much higher than the matrix, whereas particles that are formed by solid-state precipitation tend to have lower melting points than the matrix.

7 POSTSCRIPT

This concludes our tour of common engineering materials. It is clearly not comprehensive, as there is a wealth of detail available about each and every material. Indeed, some of us have spent a lifetime investigating just one or two materials! It should, however, provide you with a framework which should help you to organize what you subsequently learn about materials.

ACKNOWLEDGEMENTS

Grateful acknowledgement is made to the following sources for permission to reproduce material within this Appendix.

Figures

Figure A.3: Courtesy of the Humber Bridge Board;
Figure A.14: E.F. Boultbee and G. Schofield, *Typical Microstructures of Cast Metals*, © 1981 The Institute of British Foundrymen.

Course team acknowledgements

This appendix was prepared for the course team by Lyndon Edwards, Mark Endean, Nick Braithwaite and Stan Hiller.

Trademarks

Araldite® is a registered trademark of Vantico AG, Basel, Switzerland; and of Ciba Specialty Chemicals in the USA.